ACADEMY-INDUSTRY RELATIONSHIPS AND PARTNERSHIPS

Perspectives for Technical Communicators

Edited by

Tracy Bridgeford
University of Nebraska at Omaha

and

Kirk St.Amant
East Carolina University

Baywood's Technical Communications Series
Series Editor: CHARLES H. SIDES

Routledge
Taylor & Francis Group
LONDON AND NEW YORK

First published 2015 by Baywood Publishing Company, Inc.

Published 2017 by Routledge
2 Park Square, Milton Park, Abingdon, Oxon OX14 4RN
711 Third Avenue, New York, NY 10017, USA

Routledge is an imprint of the Taylor & Francis Group, an informa business

Library of Congress Catalog Number:
ISBN 13: 978-0-89503-906-4 (hbk)
ISBN 13: 978-0-89503-907-1 (pbk)

Library of Congress Cataloging-in-Publication Data

Academy-industry relationships and partnerships : perspectives for technical communicators / edited by Tracy Bridgeford, University of Nebraska at Omaha and Kirk St.Amant, East Carolina University.
pages cm. -- (Baywood's technical communications series)
Includes bibliographical references and index.
ISBN 978-0-89503-906-4 (cloth : alk. paper) -- ISBN 978-0-89503-907-1 (pbk. : paper) -- ISBN 978-0-89503-908-8 (e-pub) -- ISBN 978-0-89503-909-5 (e-pdf)
1. Communication of technical information--Study and teaching. 2. Technical writing--Study and teaching. 3. Academic-industrial collaboration. 4. Industrial efficiency. I. Bridgeford, Tracy, 1960- II. Bridgeford, Tracy. III. St.Amant, Kirk.
T10.5.B72 2014
601'.4--dc23

Table of Contents

SECTION III: ENGAGEMENT VENUES

SECTION IV: RESEARCH AND WRITING OPPORTUNITIES

FOREWORD

Considering Partnerships and Relationships in the Field

Carolyn Rude

For strength and sustainability as a field, technical communication requires close relationships between academia and industry. Our work toward shared goals is complementary. The most obvious connection is in job placement. Industry depends on academic classrooms to prepare qualified employees, and academia depends on industry's willingness to hire graduates in order to justify academic programs. Although academics do not define their work primarily in terms of job placement, the reality is that academic programs in technical communication would not exist without opportunities for placement of graduates. Academic programs in most or all disciplines have practice counterparts, and when academic programs flounder (often, for example, in English and other humanities departments), it is partly because the connections with practice are tenuous.

However, the academia-industry relationship in technical communication should be stronger than jobs and placement. Other academic fields with research and practice components, such as engineering and medicine, depend on academic research to improve practice, and researchers depend on practitioners to help define the research questions. This work is integral and collaborative, and that kind of relationship should be the aim for our field as well. Land grant universities, where many programs in technical communication are located, have missions to establish partnerships with industry and the community, and many have offices dedicated to developing and maintaining academy-industry partnerships. Other universities choose this mission (in his chapter, Russell Carpenter references 300 "community-engaged" universities).

In the 21st century, close academia-industry partnerships in technical communication have been local more than fieldwide, driven usually by individuals who take the initiative to make connections. Evidence for this claim is the infrequency with which articles on contemporary practice appear in the journals to which academics contribute. Ironically, academics have learned more from industry than the other way around. Several barriers seem to get in the way of partnerships.

One barrier is the perception that academy and industry interests differ in ways that preclude partnership. Some writers in this volume reflect a version of that view. Herb Smith references "theory versus practice," for example, to explain the undeveloped potential for mentoring and internships. Charlotte Robidoux, Beth Hewett, and David Overbey explore the "competing values" of academia and industry based on their respective allegiances to theory or practice. Jim Nugent and Laurence José, writing about stakeholder theory, feel unnecessarily apologetic for any inadvertent implication that they recommend a "managerial" approach to academic program development. These writers reflect something much bigger than their own views—a field's sense that academia and industry are inherently divided, not just in technical communication but in the values and cultures that mark the parts of the field as different and only uncomfortably related. Industry is profit driven, but so are universities in their own ways, and academics must respect the work of our colleagues in industry in developing information about the products we all use.

A second barrier is that the development of graduate programs beginning in the late 1980s may have had the unintended consequence of widening the divide. The focus of academics somewhat shifted from teaching for developing career technical communicators to establishing the field's research. With programs mostly located in English departments, faculty have followed the humanities model of pursuing their own research interests rather than trying to define the needs of the field and establishing the shared research focus that Steve Bernhardt recommends in this collection. These two approaches to research—driven by individual interests and driven by a shared research focus—do not have to be mutually exclusive. We can explore possible contributions of technical communication knowledge to a wide variety of communication practices through individual initiatives while at the same time embracing our responsibility to the field's needs and a shared research agenda.

Third, for various reasons, the opportunities for academics and practitioners to meet in shared forums, especially at the national conference of the Society for Technical Communication, have diminished. The journals in which academics most frequently publish, including *Technical Communication Quarterly* and the *Journal of Business and Technical Communication,* rarely feature industry contributors. Thus, the field is marked by two parallel and infrequently shared conversations about its directions and concerns. It is hard for academics to know the needs of the field and to develop research and curricula appropriate to practice if we do not talk with people in industry. At the same time, without this talk, people in industry have little basis for understanding what academics can contribute to the field other than teaching potential employees.

These challenges—a habit of thinking of binaries and differences, individually driven research projects, and lack of shared forums—may compromise academy-industry partnerships more than any inherent differences between the different expressions of the work of this field, whether in research and teaching or in practice. They point to changes that can be made. Perhaps, for example, we need to stop letting binaries frame our conversations and to foreground shared interests in such issues as intercultural communication, new media, and ethics, all of which have theory and practice components. Even when our approaches, expectations, and values differ, we have much to learn from each other to warrant the search for common ground and the new knowledge that will enhance both theory and practice. Conversations about shared interests and needs should occur at the program level and beyond to determine foci for research. And we need deliberate and intentional efforts to meet and talk in order to know more about each other and to define possibilities for partnerships.

Fortunately, people in several organizations and individuals are acting on the urgency to establish stronger relationships. The Council for Programs in Technical and Scientific Communication (CPTSC) established its 2011 conference theme as "Academy-Industry Relationships and Partnerships." In addition, the Society for Technical Communication (STC) and CPTSC collaboratively organized a preconference workshop attended by members of both groups. This volume has developed from the papers and discussion at that meeting. One good sign about this collection is that while the authors are mostly university faculty, Charlotte Robidoux from Hewlett-Packard has contributed a chapter in collaboration with two university partners; the chapter reports on a collaborative book project with two of the authors. Welcoming an industry voice into a mostly academic publication is a step toward hearing each other that should be replicated. Collaboration on publication is a great model. In another example of a professional organization taking the initiative to bring academics and industry practitioners together, the Center for Information Development Management invited four academics to its conference in September 2012 and is pursuing ways to continue dialog among various parts of the field, including collaborative publication.

Although these examples are encouraging, more needs to be done by more professional associations and individuals in the academic programs and industry. We have been talking about academy-industry relationships for at least three decades, but our talk and the establishment of goals must be accompanied by action. If efforts such as joint conferences and collaborative publication became the norm for our field, we would have a more robust and sustainable field. The stakes are high.

This collection provides some ideas for how to proceed. Stephen Bernhardt's suggestion that each of the graduate academic programs define a shared research focus can start the conversation about what our field, not just individual researchers within it, might need to know and how to find out. James Dubinsky invokes the concept of "engagement" from land-grant universities to suggest that academics build relationships with industry not with the mindset of expert problem solvers but as participants who want to help document a community's assets. He insists that we talk openly and civilly about issues. Herb Smith, Susan Katz, and Kenneth Price affirm the bridges established by internships, mentoring, and experiential learning, which not only prepare students for their lives after graduation but also benefit academic programs. Students can bring back valuable knowledge about practice from these experiences that can inform teaching and curriculum. Russell Carpenter offers details on an industry-academia-community partnership to show how a service-learning project helps partners make connections in a rural area with emerging technologies as a means of connection. All these authors offer advice about implementing such programs.

An encouraging feature of the collection as a whole is the number of detailed reports of successful academy-industry partnerships in both teaching and research. All these successes reflect the need for negotiation and flexibility regarding expectations and procedures as well as mutual respect, but in the end the partnerships accomplish goals that neither academics nor industry practitioners could accomplish on their own. The authors provide good models for academics who might be tentative about entering into partnerships with industry.

Charlotte Robidoux, Beth Hewett, and David Overbey explore the theory-practice binary, but instead of using it to dismiss the possibility of collaboration, they open up the discussion by examining the differences between theory and practice through the canons of rhetoric. Their chapter also affirms the idea that recognition of shared interests (in this case, virtual collaborative writing) is a motivation for working through divides.

Tharon Howard reports on a partnership in which an academic center provides research on usability for industry partners. Significantly, the relationship is reciprocal, and the university center changes and adapts as industry needs change from an accommodationist to a constructivist approach to usability testing or from evaluation of products to assessing users to support product innovation. Howard's chapter describes practices that build trust among partners and offers guidelines for formal arrangements to protect partners from expectations that are not feasible.

Barbara Heifferon describes a consulting situation that required a reframing of the client's plan. She demonstrates "theory in action" as she uses theory to reframe the situation. She enters her industry teaching assignment with the attitude not of expert but of partner in drawing on the local knowledge of the participants.

John Spartz and Ryan Weber expand the definition of what academy-industry partnerships might mean by exploring the relationship between technical communication and entrepreneurs. They have investigated "the types of documents functioning entrepreneurs produce, the types of communication and writing skills considered important for business founders, and the types of writing tasks entrepreneurs perform to fund, start, maintain, and grow their businesses." They project excitement at the opportunities awaiting academics who venture into this area of business writing, which should result not just in curriculum but also in contacts.

Although this collection originates in a problem to be solved and one that has resisted solution for a long time, the tone overall is optimistic. The authors are realistic about challenges, but they offer evidence of methods and attitudes to manage the challenges in the detailed examples of success. The successes depend on personal initiative, mutual respect, flexibility, and adaptability. Congratulations to the editors and authors for making visible both the need to pay attention and the evidence that efforts can result in mutually beneficial partnerships. I hope the ideas and stories in this collection will open the path for other successes to follow.

INTRODUCTION

Rethinking the Nature of Academy-Industry Partnerships and Relationships

Kirk St.Amant

Tell someone that you work in the area of "technical communication," and chances are you'll receive a curious, somewhat confused look in response. It doesn't matter if you're an academic researcher or an industry practitioner; the look is almost always the same. And we've all seen it at some point or another in our careers—it's that expression of "I kind of know what that is/what you mean, but I'm not 100% sure."

Here's where the really interesting part happens: How do you respond to the inevitable question, "So what exactly is that?" or "So what exactly do you do?" It's at the point when we respond to such questions that things really begin to differ.

Academic researchers might begin by explaining what rhetoric is and how it relates to approaches individuals use to convey technical information to different audiences. Industry practitioners, however, might respond with an overview of content-management practices or a summary of what usability (or user experience design) is. These differing responses to the same questions—the ones we all seem to get regardless of the area in which we work—reveal fundamental differences in how the field of technical communication is viewed by its members. And these differences often come down to an internal (i.e., within-the-field) binary related to how technical communicators perceive, conceptualize, and discuss the field, both among themselves and with others.

PERCEPTIONS OF THE FIELD

As a field, technical communication is often viewed as composed of two different spheres. One sphere, that of academia, is generally thought of in terms of the educational and research contexts associated with universities and colleges. The other sphere, industry, is usually seen as encompassing workplace contexts that exist outside of educational settings. Within this framework, the academic realm tends to be associated with more theoretical and abstract approaches to technical communication. Industry, in turn, is often viewed in terms of the concrete and the applied. This split perception of the field often leads to the question, *What is the nature of the relationship between these two spheres?*

Responding to this question is not easy, because it requires one to examine a range of attitudes that seem to span a spectrum of opinions. To some individuals, the perceived academy-industry difference in the field constitutes a split or a divide that is a source of tensions and disharmony. According to this perspective, members of one segment of the field often view the other with suspicion and, in some cases, contempt. To others, these two spheres represent an opportunity for collaboration across a greater expanse of ideas and approaches. For these individuals, the "other side" constitutes potential opportunity, but the question is how to recognize it.

The solution to addressing this situation involves finding the topics around which or the areas in which mutual interest can allow collaboration to take place. In each case, the issue to address is how to create effective academic-industry interactions and partnerships that maximize potential benefits and minimize prospective conflicts. The answer lies in finding models and methods that let the members of both spheres effectively establish and maintain such relationships. The answer also lies in identifying common contexts in which each side can benefit from interaction with the other, often in relation to a common cause or a mutually recognized objective. The entries in this volume represent 11 different approaches to achieving this goal.

THE ORIGINS OF THIS COLLECTION

The ideas for this volume took shape during the 2011 annual meeting (i.e., conference) of the Council for Programs in Technical and Scientific Communication (CPTSC). The theme of that year's meeting was "Academy-Industry Relationships and Partnerships," and the event represented multiple points of synergy across the field. First, a number of panels were composed of academia and industry participants who had come together to discuss approaches for fostering effective relationships between these two areas. Second, the organizers of that year's annual meeting worked with members of the Society for Technical Communication (STC)'s Academic Special Interest Group (SIG) to coordinate an STC-sponsored and STC-organized "preconference," which featured presentations and discussions involving members of academia and industry. These factors, and the discussions and the presentations that came out of these two colocated events, allowed for key conversations to take place across the academy-industry spheres. These discussions, moreover, focused on how members of all parts of the field could create relationships beneficial to *students*.

This book project emerged from a desire to capture the energy, ideas, and approaches coming from these conversations.

Several of the chapters in this collection began in the form of short (5–7 minute) and informal presentations given at the 2011 CPTSC annual meeting. Others came from the discussions that occurred in the months following this event. And each of these entries presents different ideas that continue to shape the synergistic discussions that began at that 2011 meeting. These chapters also represent a relatively broad range of approaches members of the field can use to foster academy-industry partnerships and relationships within various contexts. They include everything from internship models to community outreach practices to new perspectives on the nature of research in the field. In so doing, the entries in this collection examine such partnerships from a variety of levels—from individual activities to group/class projects to new ways of envisioning programs and curricula. The key in all of these cases is *balance*. That is, members of the field need to find ways to maximize the benefits each area can contribute to a given process (generally, an educational one) through an increased understanding of and a mutual respect for the "other" side.

THE OBJECTIVES OF THIS COLLECTION

The examination of a complex subject often starts with a series of small actions. For this reason, this text should be thought of as an overview of one piece of the greater, multifaceted puzzle that is the field of technical communication. The overarching concept behind this text is not to present a comprehensive analysis of this overall topic area. (Such a treatment would require several volumes.) Rather, the goal of this collection is to provide a sampling of approaches readers can use when considering how they, as members of academia or industry, might approach this idea of collaborating within different contexts.

The objective of this book is thus not to serve as a definitive source on how we, as members of the technical communication field, consider such relationships. Instead, readers should view this book as a tool, or a reference, individuals can use to explore or to think about such relationships in a way that best addresses the academic or the industry context of the reader. To this end, the editors encourage readers to consider how to apply, expand upon, or contribute to the ideas presented in the chapters contained here. Readers are also encouraged to use this text as a mechanism for continuing the discussion of relationships and partnerships started at the 2011 CPTSC annual meeting.

THE STRUCTURE OF THIS VOLUME

To examine various ideas and approaches to fostering academy-industry relationships, the editors have organized this collection into four sections. Each section represents a general category readers might consider when thinking about establishing different kinds of partnerships within the field. Within this context, there is no particular order in which readers should move through the four sections. Rather, each section should be viewed as a different, general approach to engaging in academy-industry relationships, and it is the reader's decision to consider which approach might most interest or best work for her or him.

That said, the four sections have, to a degree, been organized according to the idea of *contexts* or settings in which interactions might take place. (The idea being that these contexts can be fertile ground for examining such relationships.) The collection begins by looking at the educational (or the programmatic) contexts in which members of academia and industry can collaborate. From there, the examination shifts to approaches for introducing students and faculty into different industry contexts via mechanisms such as internship and mentoring programs as well as on-site industry training opportunities. The entries in the next major section looks at the role centers, labs, and outreach programs—entities often associated with making connections to the local community—can play in establishing different contexts for academy-industry relationships. The collection's final section presents ideas on how the research and writing processes can create a context in which members of academia and industry might collaborate to investigate topics of mutual interest and to share information across the field. This approach can help readers better identify the different context(s) available to them and begin to explore approaches to fostering cross-field relationships in that setting or those settings.

With these ideas in mind, let's take a look at the four sections.

Section I: Programmatic Perspectives

In essence, effective education centers around programmatic factors. That is, the value students gain from their educational experiences is closely linked to how the academic programs in which they study connect ideas and practices across different courses, options, and opportunities in an overall program. It is for this reason that academics and industry practitioners alike so often speak of education in terms of good programs (e.g., "this is a good program to apply to" or "this is a good program to have graduated from") versus an individual class or instructor. Establishing effective and sustainable academy-industry relations in an educational context therefore requires one to think of how such relationships can be forged and thought of in terms of programs. The entries in this section examine different approaches to and perspectives for facilitating academy-industry relationships on a programmatic level.

In this section's first entry, Jim Nugent and Laurence José discuss how the idea of stakeholders can provide insights into how we think about academic programs in the field, who may participate in such programs, and at what levels. In so doing, Nugent and José explain how stakeholder theory can help members of the field think of educational programs as opportunities for individuals in academia and industry to contribute to and to shape educational experiences in meaningful ways. Next, John M. Spartz and Ryan P. Weber examine how a focus on entrepreneurship, at a course and a curricular level, can help technical communication programs become contexts for collaboration and interaction across the academic and industry segments of the field. In so doing, Spartz and Weber present the results of a research project used to reconsider the nature of such relationships in terms of the teaching of technical communication as related to entrepreneurship. The section then concludes with Kenneth R. Price's review of how experiential learning projects and classes, including service-learning projects and internship experiences, can become effective mechanisms that members of academia and industry can use to collaborate via developing and using assessment practices. The successes of such relationships, Price notes, are connected to assessment practices that allow all involved parties to contribute in ways that recognize the benefits each brings to such activities.

Section II: Industry Contexts

Quite often, academics view the industry context in terms of some oft-mentioned but little understood space in which students apply the ideas covered in their classes. An understanding of that space, however, is essential to everything from developing academic programs that prepare students for life after graduation to providing funding for initiatives to examining research topics of mutual interest. Thus, the more academics can collaborate with industry partners in this space, the better they can pursue such undertakings. Education—writ broadly—can provide a range of opportunities to cultivate such relationships within an industry context. These opportunities can involve everything from formal internship programs to informal industry mentoring to conducting workshops in industry venues. In all of these cases, the key to forging relationships and partnerships involves a willingness of all parties to collaborate in an industry setting. Through such interactions, both sides (academia and industry) can learn how they might complement one another. The entries in this section examine this context (and the prospects for collaborating in it) in terms of educational activities that can bring the two sides together around a topic of mutual interest: students.

This section begins with Susan M. Katz's overview of the internship program at one particular institution, North Carolina State University (NC State), and a discussion of how internship programs such as this one can help academics and students gain a better understanding of industry contexts. In presenting the different aspects of NC State's internship program, Katz also provides ideas and frameworks readers can use to create new internship programs or modify existing ones to better facilitate discussions, collaborations, and relationships across academic and industry spheres. Herb J. Smith's entry expands upon this idea of internships as a venue for academy-industry partnering by providing a rubric for fostering the relationships discussed by Katz in her chapter. In so doing, Smith discusses pairing internships with mentoring activities (i.e., pairing students with a mentor working in industry) to further develop such relationships by providing other methods for understanding the industry context. In the section's final entry, Barbara A. Heifferon discusses how academics can enter into and contribute meaningfully to industry contexts in the form of industry training sessions. Heifferon's chapter, which presents her experiences developing and delivering writing training in an industry setting, provides academics with an example of how they can use teaching practices as a method of engaging with and collaborating with industry colleagues in settings outside of the conventional university, college, or community college classroom.

Section III: Engagement Venues

Educational, government, and (sometimes) business institutions often house entities that focus on outreach to and collaboration with the local community or the greater regional area. These entities often exist in the form of centers, labs, focused programs, or offices that offer individuals, generally students or volunteers, different approaches to and opportunities for engaging with partners from academia and industry. In some instances, these units send students out into the community to apply what they have learned in their classes in different local settings. In others, they invite members of industry into academic environments via projects with which students can engage. In almost all of these cases, the objective is the application of classroom-derived knowledge to extracurricular projects. And through such processes, created by academy-industry partnerships, participants can apply ideas within broader social contexts. This section examines the kinds of contexts associated with such partnerships.

In the section's lead entry, Russell G. Carpenter analyzes a partnership that linked the Noel Studio for Academic Creativity at Eastern Kentucky University (EKU) with a middle school and the Bluegrass Army Depot (BGAD) in Richmond, Kentucky. In so doing, Carpenter uses a humanistic lens to examine how emerging technologies can facilitate academy-industry partnerships across different regions, including rural areas where forging such partnerships can be more difficult. Tharon Howard's entry, in turn, provides a retrospective of the academic-industry partnerships that emerged from a usability testing facility/lab over the course of almost two decades. In reviewing the evolution of such relationships, Howard describes how his lab's focus changed over time to address the new contexts in which industry partners were operating and how managing such change can help develop fruitful collaborations across academia and industry. James M. Dubinsky then concludes this section with a discussion of how community-based engagement projects can provide a forum that members of academia and local industry can use to build community capacity and encourage creative leadership. In discussing this project, Dubinsky reveals how understanding the power of "place" (i.e., local context), in combination with an engagement approach, can engender academy-industry relationships that meet the technical, social, and human challenges of a community.

Section IV: Research and Writing Opportunities

In many ways, the topic of research can be a central area of synergy for academy-industry partnerships. In these relationships, academics bring new theories and models to test and apply, and industry practitioners bring new materials for analysis or new questions that need answers via research. Similarly, the process of sharing results and ideas through collaborative writing can facilitate relationships based on mutual interest in examining a common topic. In both instances, success involves finding a common interest and then identifying—and appreciating—what the "other side" can contribute to the study of that shared interest. Through such recognition of common goals and differing contributions, individuals in academia and industry can examine topics and explore ideas in a more complete and holistic way. The entries in this section examine approaches to such relationships as they relate to research (information seeking) and writing (information sharing) practices.

In the section's lead entry, Stephen A. Bernhardt examines approaches to setting programmatic research agendas that have been successful. In so doing, he identifies ways academic programs can define and engage in research activities that benefit departments and are valued in various industry contexts. Bernhardt's entry thus brings the discussion back to the collection's initial examination of programmatic contexts and helps readers reconsider such contexts from the perspective of shared research interests. The question then becomes what activities can both sides undertake to pursue this kind of collaboration? One answer is provided in the section's second entry by Charlotte A. Robidoux, Beth L. Hewett, and David W. Overbey. These authors analyze the experience of using online media to coordinate a collaborative writing project that involved 22 writers affiliated with either academia or industry. The collaborative process described by the authors provides an example of how individuals from academia and industry can work together to develop and share information on a topic of mutual interest.

FINAL THOUGHTS

There are numerous ways to create and to maintain effective academy-industry partnerships and relationships. The entries in this collection represent but a sampling of such approaches. The next step is for individuals to consider how they might employ or build upon the approaches presented here—or even contribute new models for interaction. The editors thus encourage readers to view these chapters as prompts for thinking about the nature of these relationships and to do so in ways that can foster further discussion across the field. The editors also encourage readers to consider how they might combine the ideas covered in this collection with their (the readers') own experiences in order to contribute to an ongoing examination of such ideas within the field. After all, the key to successful relationships is conversation, and the more we can talk with each other—across the different spheres of the overall field—the better all members of the field can move forward more effectively and more evenly.

And so the conversation begins.

SECTION I

Programmatic Perspectives

CHAPTER 1

Stakeholder Theory and Technical Communication Academic Programs

Jim Nugent and Laurence José

ABSTRACT

The term "stakeholder" is ubiquitous within the disciplinary parlance of technical communication, yet often its usages are generic, its meanings unclear, and its intellectual origins unacknowledged. Meanwhile, stakeholder theory stands as one of the central and most influential theories in business ethics, and its sizable literature spans over three decades of scholarship across numerous academic disciplines. In this chapter, the authors explore the potential of stakeholder theory for revising our understanding of technical communication academic programs and the academy-industry binary. They suggest how the theory might inform program administration, including identifying and classifying stakeholders, reconciling the competing interests of those stakeholders, and remaining aware of the ethical consequences of our programs' work.

INTRODUCTION

As a theoretical discipline rooted in workplace practice, technical communication finds itself circumscribed by a persistent dichotomy between academy and industry. Hayhoe (2004) observed the "misunderstanding, distrust, and sometimes downright animosity that exists between those who teach technical communication and those who are technical communicators in industry" (p. 101), even going so far as to characterize the division between these perspectives as a militarized border with "barricades" and "barbed-wire fences" (pp. 102, 108).

Much of our understanding of the work of technical communication and its academic programs is predicated on this dichotomy, and the antagonism suggested by Hayhoe's metaphors is often taken as a given in our discourse.

Over the decades, technical communication scholars have presented various frameworks for theorizing, contending with, and even transcending the academy-industry dichotomy. For instance, Carolyn Miller (1989) suggested Aristotelian *praxis* as a middle term between the realms of academy and industry, theory and practice, the ivory tower and the marketplace, and so on. Likewise, Thomas Miller (1991) turned to classical rhetoric to define technical communication as a form of social *praxis* and to avoid remaining "marginal members of the social worlds that we work with inside and outside the academy" (p. 71). Johnson-Eilola (1996) warned against technical communication following a "support model" or "service orientation," suggesting a postindustrial model of technical communication as "symbolic-analytic work" to serve as a productive middle ground. And Kline and Barker (2012) have attributed no small significance to overcoming this dichotomy by asserting that the realization of "professionalism in technical communication depends on bridging the gap between academics and practitioners" (p. 34).

In the present chapter, we contend that the concept of stakeholders can provide both theoretical and practical insights into the long-standing dynamic between academy and industry in technical communication. The term "stakeholder" has enjoyed almost faddish success in popular discourse over the past decades as a way to refer to those parties poised to affect, or be affected by, the actions of an organization. As we discuss in this chapter, however, a surprisingly wide-ranging body of scholarship lies behind the term's buzzword façade: *Stakeholder theory* stands large within business ethics, management, and a host of other academic disciplines, and it offers a number of insights into the effective and ethical administration of organizations. We suggest that by thinking about technical communication academic programs less in terms of their dichotomous opposition to industry and more in terms of their polychotomous relationship to diverse stakeholders, we may come to see our curricula in ways that neither presume subservience to industry nor embrace academic irrelevance.

We begin this chapter by illustrating the inadequacy of the academy-industry binary as a conceptual lens, using a brief example from a professional writing course. We then try to move the stakeholder concept onto better defined intellectual ground by drawing from stakeholder theory scholarship. It is impossible for us to adequately represent the many decades of work in this field here, but we feel it is important to draw attention to the intellectual origins of the seemingly ubiquitous and well-understood term "stakeholder," as those origins are rarely acknowledged in the scholarship of technical communication. Even more importantly, however, we believe that stakeholder theory can offer us positive recommendations for program administration, including frameworks for identifying and classifying their constituent stakeholders, reconciling the competing interests of those stakeholders, and remaining aware of the ethical consequences of our programs' work.

TRANSCENDING THE ACADEMY-INDUSTRY DICHOTOMY

Before we turn to the intellectual origins of stakeholder theory, we first offer a brief example that illustrates the limits of the academy-industry dichotomy and suggests the need for alternate frameworks to understand the work of our academic programs. In this example, an introductory professional writing classroom became a contact point for teachers, students, and professionals and demonstrated that the positions represented there were too fluid, transient, and highly localized to be completely accounted for by a strict academy-industry dichotomy.

The Dichotomy in the Classroom

In spring 2011, Laurence taught two sections of WRT 200: Introduction to Professional Writing, a core course in the independent writing program at Grand Valley State University (GVSU) in Allendale, Michigan. In addition to primarily serving writing majors, this course serves secondary populations of writing minors and nonmajors. Near the end of the term, Laurence invited two professionals into the classroom to discuss their experiences. The decision to initiate a dialogue between students and professionals was directly motivated by a desire to somehow confront the academy-industry dichotomy in her class. The goal was to help students establish connections between what they had learned in the classroom and the practices and responsibilities of professionals in the workplace.

If the tensions between the industry and the academy are characteristic of the field as a whole (St.Amant & Nahrwold, 2007), these tensions are felt particularly within introductory courses. Indeed, introductory courses not only lay the foundation for students to understand the knowledges comprising the discipline, but they also act as windows into the possibilities for, and applications of, the practices learned in the classroom. At GVSU, for instance, WRT 200 has a dual function: It introduces students to the main genres of professional writing (the practice), while also emphasizing the disciplinary nature of professional writing as a field (the theory). Like many introductory courses, WRT 200 is a privileged location to examine, to refine, and to interrogate the academy-industry dichotomy.

The two professionals who visited the course worked respectively as a senior sales writer at a local insurance company and as a director of marketing and communications at the state branch of a major nationwide nonprofit. In addition to their full-time jobs, they also regularly engaged in freelance work. The professionals briefly demonstrated their freelance writing websites before, discussing their educational background, their job responsibilities, and their general work practices.

Not surprisingly, two main topics seemed to preoccupy the students: professional activities (i.e., what the speakers' work consisted of, what their schedule was like, what was challenging, what design software they used the most, etc.) and the relationship between the classroom content and the actual contingencies faced by professionals. In other words, students were as much interested in learning about the daily lives of the professionals as they were about the connections between their current practices as students and their practices as future members of the industry.

This classroom conversation quickly highlighted the limits of a strict and static academy-industry dichotomy: Although the actors present that day apparently belonged to two clearly demarcated camps, the dialogue disrupted these categories and unveiled a more complex dynamic. For instance, both professionals were GVSU alumni, and one of them was himself an alumnus of the writing program. Once the professionals began to address students' questions regarding the relevance of classroom practices for professional practice, the duality of their perspectives became apparent: They were speaking both as (current) professionals and as (former) students. The fact that they had both graduated from the university that their student audience was attending emphasized the fluidity of their categorical positioning. Their voices acquired an insider's dimension as they were explicitly speaking as former members of the academic context.

Viewing the speakers simply as self-interested agents from industry seeking to influence the formation of potential future hires would do little justice to the complexity of their positions; reducing the dialogue to "us versus them" would have considerably limited the significance of the exchange. Similarly, on hearing the questions, it was evident that student perspectives could not simply be reduced to their positions as members of a professional writing course during a specific semester. Students' questions and priorities were dependent not only on their chosen degree programs and professional aspirations but also on their extra-curricular experiences or, perhaps in some cases, on their desire to simply do well in the course. In this context, the multiplicity of the students' perspectives could not be read solely through the filters of "student position" or "academic context." Their positions too were fluid, transient, and highly localized.

Recognizing Stakeholder Dynamism

What we take from this brief illustration is not that the academy-industry binary does not exist; rather, we seek to emphasize the dichotomy's severe limitations as a conceptual lens for describing the different agents and entities that compose such pedagogical and institutional spaces. Although the academy-industry binary prompted the encounter between professionals and students, the binary is clearly not a sufficient framework for identifying and characterizing the complex interests of the different parties involved. These stakeholders do not all fit tidily into the *either/or* categories of academy and industry, suggesting the need for alternate conceptual frameworks.

SITUATING AND REFINING "STAKEHOLDERS"

This section explores the stakeholder concept and provides a brief introduction into the broader work of stakeholder theory. The term "stakeholder" appears frequently and fleetingly in technical communication scholarship on communication practice, usability, ethics, academic programs, assessment, internships, professionalization, and so on (e.g., Carter, Anson, & Miller, 2003; Conklin, 2008; Coppola, 1997, 2011, 2012; Dayton, 2002; Eubanks & Abbott, 2003; Grabill & Simmons, 1998; Morgan, 1997; Salvo & Ren, 2007; Savage, 1996; Scott, 2004; Wallace et al., 2006). It also enjoys fruitful applications in the classroom: Many of us find the stakeholder concept indispensable for teaching students how to assess their particular audiences, purposes, and contexts; how to invent appropriate strategies for their rhetorical situations; and how to remain attentive to the ethical dimensions of their work (e.g., Anderson, 2010; Hartung, 1998; Zoetewey & Staggers, 2004). The stakeholder concept is alluringly intuitive, accessible, and useful, and it is with good reason that the term finds itself woven into the vocabulary of the discipline and profession.

But while the term "stakeholder" is prevalent in our disciplinary parlance, its usage has typically remained generic, its origins obscure, and its exact meanings unspecified. Within the scholarship of stakeholder theory, however, we find a more precise and well-developed articulation of the concept. By looking to stakeholder theory as articulated in business ethics, management, and other academic fields, we hope to locate the stakeholder concept within an identifiable intellectual tradition and to provide greater specificity to its definition and application. While we do not wish to suggest that academic programs are business enterprises—and we certainly do not wish to suggest that academic institutions are in need of a more businesslike approach to their administration—we believe that stakeholder theory may offer a number of conceptual insights and practical recommendations for program administration. Although scholars have employed a generic stakeholder concept to describe academic programs before, they have not done so from the perspective of stakeholder theory proper, nor have they taken advantage of the theory's unique insights into how to identify and classify stakeholders, how to balance the competing interests of those stakeholders, and how to remain aware of the ethical consequences of an organization's actions.

Origins and Intellectual History

The scholarship of stakeholder theory was set in motion some 30 years ago by the philosopher and professor of business R. Edward Freeman. Stakeholder theory holds as its central tenet that organizations are ethically responsible to more than just the narrow interests of their owners or shareholders. Rather, organizations are obligated to a broad array of stakeholders, which Freeman's 1984 volume *Strategic Management: A Stakeholder Approach* defined as "any group or individual who can affect or is affected by the achievement of the organization's objectives" (p. 46). From this fairly straightforward definition grew a sizable body of literature in stakeholder theory, which has come to be regarded as "one of the most influential theories in business ethics" (Egels-Zandén & Sandberg, 2010, p. 35).

Regarding the word "stakeholder" itself, *The Oxford English Dictionary* (2004) has cited its use as early as 1708, meaning an independent, disinterested party that holds the stakes of a bet, as is the case in pari-mutual wagering. Only in recent decades, however, has the word come to signify a party affected by or interested in a particular action. (Interestingly, the word is an autoantonym, meaning in its contemporary sense a party *with* a "horse in the race," while in its historical sense it was, often quite literally, someone *without*.) The first use of "stakeholder" as a deliberate play on the word "shareholder" or "stockholder" appears to have occurred in a 1963 memo at the Stanford Research Institute (SRI). In this memo, the term was defined as "those groups without whose support the organization would cease to exist" (as cited in Freeman & Reed, 1983). From the SRI, the stakeholder concept took hold during the 1960s and 1970s within the fields of strategic planning, systems theory, corporate social responsibility, and organizational theory (Freeman, Harrison, Wicks, Parmar, & de Colle, 2010; cf. Laplume, Sonpar, & Litz, 2008, p. 1156).

According to Freeman's account of the genesis of stakeholder theory, the term "stakeholder" was "very much in the air" during his tenure at the Wharton School at the University of Pennsylvania in the late 1970s (Freeman et al., 2010, p. 52). While the management scholars there embraced a basic stakeholder notion, he noticed that they were reluctant to engage with "issues of values, ethics, or justice" (Freeman et al., 2010, p. 51). Seeking rather to "say something about normative and justice issues" (Freeman et al., 2010, p. 51), Freeman expanded the concept of a "stakeholder" away from the SRI's strictly survival-based definition toward one that encompassed those entities who are positioned to affect an organization as well as those entities affected *by* it. He also sought "to integrate ethical and social issues directly into strategy" (Freeman et al., 2010, p. 58) and to develop the stakeholder concept into a workable theory of management.

In 1984, Freeman articulated these ideas in his seminal book *Strategic Management: A Stakeholder Approach*. Put simply, Freeman's work articulates a theory of "stakeholders" as a means to identify and map out the role of the different entities constitutive of, and affected by, an organization; to understand the holistic functioning of an organization by gauging the significance of its different constituent entities; and to call attention to the ethical ramifications of these different entities' actions for the values and the very identity of an organization. The end goal of stakeholder theory is to lead organizations to be more ethically responsible actors by compelling them to account for their effects on other parties and to remain aware of their reciprocal relationship to those parties. As Freeman and other stakeholder theorists maintain, the theory not only provides a framework for insuring the ethical administration of an organization, but it also provides a framework for insuring its practical success.

A Theory with Cross-Disciplinary Applications

Although its origins lie closest to the fields of management and business ethics, stakeholder theory has proven to be immensely versatile in application. It has been used by numerous scholars in diverse fields of study, including finance (e.g., Bae, Kang, & Wang, 2011; Barton, Hill, & Sundaram, 1989), energy policy (e.g., Alberts, 2007), risk analysis (e.g., Beierle, 2002), natural resource management (e.g., Billgren & Holmén, 2008), project management (e.g., Bourne & Walker, 2005), negotiation and applied psychology (e.g., Campbell & Mark, 2006), political science (e.g., de Bussy, Watson, Pitt, & Ewing, 2000), construction management (e.g., Fraser & Zhu, 2008), public relations (e.g., de Bussy, Watson, Pitt, & Ewing, 2000), communication management (e.g., Ihlen & Berntzen, 2007), legal studies education (e.g., Jennings, 1998), accounting (e.g., Orij, 2010; Roberts & Mahoney, 2004), corporate governance (e.g., Mygind, 2009; Yamak & Süer, 2005), health care (e.g., Werhane, 2000), environmental development and sustainability (e.g., Wang, 2010), and others.

In addition to its breadth of application, the stakeholder literature is amenable to a wide variety of interpretative approaches. For instance, scholars have interpreted stakeholder theory through the conceptual lenses of Aristotelian ethics (Wijnberg, 2000), feminism (Wicks, Gilbert, & Freeman, 1994), Jürgen Habermas' theory of communicative action (Zakhem, 2008), and *The Bhagavad Gîtâ* (Parekh, 2008). Stakeholder theorists have also expanded upon the original theory by proposing that even nonhuman actors such as the natural environment (Orts & Strudler, 2002), the state (Yamak & Süer, 2005), trees (Starik, 1995), infrastructure (Luoma-aho & Paloviita, 2010), and God (Schwartz, 2006) can stand as potential stakeholders.

The adaptability and flexibility of stakeholder theory are widely seen as its intellectual strengths (Philips, Freeman, & Wicks, 2003). As Freeman et al. (2010) have maintained, the stakeholder framework represents less a static, singular theory than it does an entire genre that can find fruitful applications within many different contexts:

> That is, rather than being a specific theory used for one purpose (e.g., resource dependence theory in management), seeing stakeholder theory as a "genre" is to recognize the value of the variety of uses one can make of this set of ideas. There is enough commonality across these uses to see them

> as part of the same genre, but enough diversity to allow them to function in an array of settings and serve different purposes. (p. 64)

To them, stakeholder theory lies squarely within the tradition of philosophical pragmatism and is part of a "Rortian and Deweyan vision of inquiry" (p. 72). This vision is concerned less with the question "'what is true' than 'how should we live,' or better still, 'how does this narrative allow us to live,' or 'what does this way of talking allow us to do'" (Freeman, 1994, p. 418).

It is in this same spirit of pragmatism and adaptation that we have turned to stakeholder theory in the present chapter. At first glance, stakeholder theory may seem ill-suited to the discipline of technical communication. After all, the theory has emerged from business-related fields and has been framed as an explicitly "managerial" theory. The theory may not sit well with many academics who, as Rude (2009) characterized, may "resist association with the power structure of science, technology, and business, which, like any dominant power, marginalizes some people and values" (p. 189). However, our particular application is intended to be appropriative in spirit and to remain consistent with the tradition of diverse and flexible adaptation that stakeholder theory has already enjoyed. Our use of this theory is not meant to be strictly literal—again, we are in no way suggesting that academic programs are businesses or that universities need to be run in a more "managerial" way. However, we feel that stakeholder theory as it has been articulated across several decades of interdisciplinary scholarship may offer us a number of practical and theoretical insights that existing applications of the stakeholder concept in technical communication have not, including ways of rethinking the academy-industry binary altogether.

STAKEHOLDER THEORY AND TECHNICAL COMMUNICATION PROGRAMS

Complicating Program Stakeholders

The concept of *stakeholders* has been present in the technical communication literature for over 20 years. Yet, as we will show, when the concept has been used to describe programmatic contexts, it usually has remained generic in its application and has failed to exploit the potential of stakeholder theory proper to account for the complex interplay among academic programs and their stakeholders (classified as "industry," "academy," or otherwise). Among other things, stakeholder theory is able to provide tools for identifying and classifying organizational stakeholders, for balancing the competing interests of those stakeholders, and for remaining aware of the ethical consequences of an organization's work. Existing applications of the stakeholder concept to technical communication programs have performed adequately on the first count—stakeholder identification—but we believe that stakeholder theory can offer additional insights into how to discern where stakeholders stand strategically in relation to the program (stakeholder analysis) and how to determine the degree to which various stakeholders merit a program's attention (that is, their relative salience).

The central admonition of stakeholder theory is that organizations are obligated to take into account the voices and viewpoints of those parties poised to affect (or be affected by) the organization's actions. To meet this obligation, of course, organizations must know who their stakeholders are. The literature of stakeholder theory is rife with various frameworks and procedures for performing such identification. Bryson (2003), for instance, catalogued at least 15 different methods of varying complexity. The first and simplest of these is to brainstorm all of the groups poised to affect (or be affected by) the organization's actions and to then imagine what constitutes the criteria for the organization's success or failure in the eyes of those stakeholders. More in-depth procedures entail drafting "power versus interest" grids, "stakeholder influence diagrams," "participation planning" matrices, "bases of power/directions of interest" diagrams, "stakeholder-issue interrelationship" diagrams, and other strategic documents.

Beyond stakeholder identification, Ackermann and Eden (2011) noted that "who stakeholders are is related to the multifarious nature of the demands they can make on the organization" (p. 179). As such, effective stakeholder analysis also calls for some assessment of where stakeholders stand relative to the organization. Sorting stakeholders into primary and secondary categories is one popular strategy for doing this (Clarkson, 1995). Others have proposed sorting interested parties into the more finely grained categories of stakeholders, stakewatchers, and stakekeepers (Fassin, 2010); normative and derivative stakeholders (Phillips, 2003); generic and specific stakeholders (Carroll, 1989); and so on. But regardless of the mechanism employed, stakeholder analysis aims overall to "evaluate and understand stakeholders from the perspective of an organization" and to interrogate "the position, interest, influence, interrelations, networks and other characteristics of stakeholders, with reference to their past, present positions and future potential" (Brugha & Varvasovszky, 2000, p. 239).

Within the literature of technical communication, a few commentators have performed a sort of stakeholder analysis of academic programs. But as the examples below show, these analyses do not go very far beyond basic stakeholder identification. For instance, Anderson (1995) observed that a technical communication

> program's academic stakeholders include its students, graduates, and faculty, plus the department that houses the program and the institutional administrators who provide support and sustenance for it. Academic stakeholders also include the academic discipline of technical communication (represented, for example, by the Council for Programs in Technical and Scientific Communication) . . . . On the professional side, a program's stakeholders

> include not only past, present, and future sponsors of its interns and employers of its graduates, but also the profession of technical communication (represented, for example, by STC), because its continued health depends on the quality of the academic programs that educate the profession's future members and leaders. (p. 630)

Meanwhile, Carnegie (2007) acknowledged the power of certain program stakeholders to determine program resources:

> It is the political and bureaucratic stakeholders who decide what resources and how many will be allocated to institutions and their programs. The U.S. federal government, for example, provides access to student financial assistance programs only to institutions and programs it deems eligible. State governments must decide how to allocate funding to numerous universities and community colleges. Higher education institutions, in turn, decide how to allocate funding and support to different colleges, departments, and programs. (p. 450)

Rude (2009) identified further stakeholders as "faculty within technical communication programs, faculty in other programs within the university who compete for resources and the right to determine the nature and value of knowledge, organizations that hire graduates, and students themselves" (p. 197). Eubanks and Abbot (2003) asserted the need for academic programs to remain attentive to their stakeholders and to make curricular changes as needed, "just as a practitioner would do with interim usability testing" (p. 39). And St.Amant (2003) advocated a "stakeholder education approach" to internships, one that recognizes that there are many different entities having varied stakes in our internship programs.

Although quite instructive, these applications of a generic stakeholder concept are not as systematic as the methods described in the literature of stakeholder theory. In particular, by neglecting to assess where stakeholders stand relative to academic programs or to assess the relative legitimacy of those stakeholders, these analyses do not address one of the most crucial insights of stakeholder theory proper: the notion of stakeholder salience. While stakeholder theory suggests that all stakeholders should receive fair and respectful consideration, it does not state that every stakeholder interest merits equal status or equal shares of an organization's attention. Instead, as Cragg (2002) noted, it requires "that the *legitimate concerns* of all stakeholders be equitably addressed" (p. 138; emphasis added). Stakeholder theory does not demand acquiescence to every possible party that might affect or be affected by a decision, then. Rather, it demands a full and fair consideration of stakeholder interests and, to the extent that they are salient, an attempt to negotiate among those interests. Salience itself is a highly variable and dynamic attribute (de Bakker & den Hond, 2008; Mitchell, Agle, & Wood, 1997), and determining how to assess the salience is one of the central challenges of stakeholder theory. In fact, research on salience comprises almost one in five articles in the vast stakeholder theory literature (Laplume et al., 2008).

The notion of salience is crucial to a stakeholder theoretical understanding of technical communication academic programs, since it presents a subtle but important caveat not offered by existing applications of a generic stakeholder concept. To recognize technical communication programs as engaging in, or even being constituted by, a process of negotiation among interested stakeholders does not necessitate that our programs "sell out" to those stakeholders. Rather, stakeholder theory requires an attempt to balance those stakeholder interests with respect to their varied degrees of assessed salience. This may even mean setting aside the interests of certain stakeholders altogether; Freeman et al (2010) observed that achieving "balance between allowing external influences (stakeholders) to determine completely organizational strategies and totally ignoring them is at the heart of modern stakeholder theory" (p. 89). So in contrast to some characterizations (e.g., Gallagher, 2011), a stakeholder theoretical approach does not require program directors to attend to any stakeholders deemed illegitimate or irrelevant. Rather, a stakeholder theoretical perspective upholds a program's autonomy while also drawing attention to its ideal and material interconnectedness with a range of diverse interests.

A Hypothetical Illustration

We would now like to offer an elementary, hypothetical situation in order to illustrate some of the insights of stakeholder theory when applied to technical communication academic programs. Suppose a technical communication baccalaureate program was being petitioned by a large local employer to provide more direct instruction in digital tools such as RoboHelp within their curriculum. To address this issue successfully, according to stakeholder theory, the program should attempt to identify its relevant stakeholders, to perform an assessment of their relative salience, and to strike a balance among their varied interests in order to achieve the best possible outcome for all. To be sure, stakeholder theory itself cannot not furnish a pat or easy answer to this scenario. Stakeholder theory makes no claim to offer "an algorithm for day-to-day . . . decision-making" or "to dictate specific action in the abstract," nor does it suggest that there is only one right and proper course of action in any given context (Philips et al., 2003, p. 485). It does however offer a framework for approaching the problem, defining its scope, and accounting for its strategic and ethical possibilities.

To begin, the program should engage in a process of stakeholder identification, accounting for all possible stakeholder interests, evaluating their strategic and ethical relationship to the program, and assessing their relative salience. Such stakeholder interests could include

- The faculty's intellectual fealty to the discipline of technical communication, to the millennia-old enterprise of academia, to the traditions of critical thought, and to the respective values and ideologies of each;
- The faculty's need to maintain disciplinary and institutional *ethos* within the context of their department, school, and university;
- The long-term professional interests of students to learn theories that will serve them throughout life;
- The short-term professional interests of students to learn skills that will make them marketable;
- The program's academic reputation among its peer institutions;
- The program's reputation for providing economic opportunity for its students;
- The institution's ability to provide the necessary resources for the requested instruction;
- The institution's myriad bureaucratic, legislative, and administrative circumstances;
- The founders, benefactors, donors, and tuition payers of the institution—not to mention the citizens whose tax revenue directly and indirectly supports higher learning—and their expectations that the institution will expend resources effectively and responsibly;
- Explicit curricular standards as established by academic and professional communities such as the Council for Programs in Technical and Scientific Communication (CPTSC), the Association of Teachers of Technical Writing (ATTW), the Society for Technical Communication (STC), and so on; and
- Implicit curricular standards as established by prevailing practice.

Within this scenario, there could be many possible valid outcomes using a stakeholder theoretical approach depending on how the salience of the varied stakeholder interests is assessed in context and how balance is sought among them. For instance, it could be plausibly argued that students' stake in obtaining theoretical knowledges that will outlast the latest version of RoboHelp should trump the local employer's stake in having a supply of immediately productive employees. Or it could be argued that prevailing curricular standards calling for education in digital tools should trump the technical communication faculty's need to avoid the appearance of "vocationalism" in the eyes of, say, the literature faculty they compete with for departmental resources. Or, as is all too likely in the world outside this fictional scenario, decreasing levels of university funding could make it impossible to obtain the necessary software and hardware to teach RoboHelp, rendering the entire issue moot. Regardless of the action ultimately taken by the program, stakeholder theory provides a way to map out the complex interplay among these varied interests. More broadly, it draws attention to the ways in which our work in the classroom, the enterprise of our academic programs (and even the contours of our scholarly discipline itself) are shaped by negotiations among many interested parties.

Stakeholder theory cannot determine, *a priori*, who the "winners" will be in such negotiations; different courses of action will affect various stakeholders differently and have different affects on the instrumental success of the program. But in its application to technical communication programs, the theory does lead to a number of important conclusions:

- Stakeholder theory contends that the ideal course of action is to create the best possible outcome for as many salient stakeholders as possible. It holds that this will in turn lead to the instrumental success of the program, if not become the very measure of it.
- At the end of the day, the technical communication program remains in control. It identifies who the stakeholders are and assesses their salience. Hopefully it does so fairly, equitably, and completely. But to recognize a program's connections to multiple stakeholders does not demand acquiescence to them.
- Even a rudimentary listing of program stakeholders as we have provided above reveals that they are far more numerous and intricately connected than such a simple scenario would first suggest. Most significantly, these stakeholders do not all fit tidily into either/or categories of academy and industry, reminding us that there are serious conceptual limitations to such a dichotomy.

In sum, we feel that a stakeholder perspective on technical communication programs has the potential to be simultaneously descriptive, prescriptive, and instrumental (see Donaldson & Preston, 1995). It is descriptive in that it can successfully account for the range of interests in our academic programs, make sense of stakeholders' competing claims, and describe the political and material constitution of the program itself. It is prescriptive in its injunction to attend to salient stakeholders and to maximize positive outcomes for them all. And finally, a stakeholder approach is instrumental in that it suggests that success for the program will attend success for its stakeholders.

STAKEHOLDER ETHICS

Reconciling the Instrumental and the Ethical

Beyond its potential to serve as a way for charting the complex political and material interests that intersect with and negotiate at the locations of our academic programs, stakeholder theory holds promise in addressing a number of other issues in the field. It may help us to better understand the local contingencies of programs and why technical communication academic programs demonstrate notable variation in their curricula from institution to institution (Harner & Rich, 2005; Nugent, 2010). The theory may help us to understand projects such as the STC's Body of Knowledge initiative in ways that appreciate the breadth of interests represented within it and the great challenge of establishing balance among them. And, as we have discussed, it may help us to complicate the restrictive and inaccurate dichotomy of "academy versus industry."

However, there may be an additional reason to consider a stakeholder theoretical perspective for thinking about technical communication programs. Beyond the theory's descriptive, prescriptive, and instrumental dimensions, stakeholder theory also has a central ethical dimension. For an organization to take account of those entities poised to affect or be affected by its actions is not only a useful strategic act, but it is also an ethical undertaking: It is the first step toward recognizing that "stakeholders have names and faces and children" (Freeman, 2010, p. 9). More broadly, stakeholder theory has been articulated against what Freeman has termed the "separation thesis," or the prevailing notion that decisions about how an organization is run can be separated from ethics (Freeman et al., 2010). As is often the case in the business world, corporate decisions are made largely with the instrumental success of the organization in mind; such decisions are often classed as "business decisions" and are presumed to be immune to ethical consideration. For instance, the separation thesis is at work when certain organizational ends (such as boosting shareholder value) are held to justify their means (such as initiating mass layoffs) in what is touted as a strictly "business decision." Such an action is presumed to be an inevitable matter of fiduciary duty, "dollars and cents," or "what's best for the institution." Stakeholder theory, however, rejects the separation thesis and holds that the administration of an organization is utterly indivisible from ethics. Applied to academic programs, this central ethical claim exhorts us to remain mindful of the impact that our programs' work has on our varied stakeholders.

Stakeholder Ethics *in Vivo*

It is easy from within the academy to presume ethical rectitude on our own behalf, but some recent prominent failings have suggested that a version of the separation fallacy exists within academic contexts. For instance, in August 2011, a number of graduates of the New York Law School filed a class action lawsuit against the school contending that it had marketed its programs using misleading statistics about career and salary prospects for its graduates. Although the suit was eventually dismissed, it became clear during the trial that the school had cherry-picked survey data to make its degrees appear more beneficial to students than they really were. In this case, the institution prioritized its own short-term instrumental success over the interests of its students (an especially salient stakeholder group for any school). School administrators in all likelihood saw nothing wrong with their actions because although the data were misleading, they were not fabricated or illegal. The school seemingly engaged in a version of the separation fallacy by presuming that their students were free to make their own choice over whether to pursue a degree or not; because students were not strictly lied to or coerced and because it would lead to certain instrumental outcomes for the institution, the school's actions were no doubt conceived of as mere "business decisions."

Similarly, the University of Phoenix has been cited for poor graduation rates, for high levels of debt among former and present students, for excessive focus on profit at the expense of academic integrity, and for violating laws prohibiting compensation to school recruiters based on the number of students they enroll. These transgressions can be seen as a failure to attend to a host of different stakeholders: taxpayers (by taking federal financial aid dollars without maintaining acceptable graduation and retention rates), students (by recruiting underqualified students and allowing them to mount up irrevocable debt with little prospect of completing their degree), academic disciplines (by straying from the intellectual traditions of quality and rigor), citizens (by violating the laws that represent their interests), and so on. As a publicly traded company, the University of Phoenix is beholden to its shareholders; as such, the school's failures could be construed as an inability to establish a successful balance between their shareholder's financial interests and the interests of other vital stakeholders. In this case, the separation fallacy was manifested quite literally as a series of "business decisions."

More generally, a number of academic programs (and even entire disciplines) have encouraged overproduction of degree holders in their fields, often with disastrous career and financial consequences for students. From a stakeholder theoretical perspective, such programs can be seen as having unduly placed one set of stakeholder interests (specifically, the interests of their disciplines, affiliated academic organizations, and faculty in maintaining instrumental power and institutional longevity) over others (the short- and long-term well-being of their students). We are definitely not suggesting that a program's societal value is measured by cost-benefit ratios, employment statistics, or other strictly economic means, or that the interests and values of academic disciplines are subservient to economic considerations. But many academic programs seem to hold "caveat emptor" as a moral dictum and presume that students' material well-being after graduation is utterly irrelevant to their work. In this way, much like the separation thesis operates within business contexts, decisions within academic contexts are often justified in terms of "what's good for the institution" rather than in terms of "what's good for stakeholders." Although most program actions do not lead to ethical outcomes as stark as the examples cited above, we believe that stakeholder theory's injunction to maximize positive outcomes for all relevant parties can prevent excessive focus on narrow goals such as institutional self-preservation and can help academic programs to remain not only ethical but also relevant and effective.

CONCLUSION

To be sure, stakeholder theory is not poised to obliterate "academy" and "industry" as meaningful categories, nor can it provide easy solutions for conflicts that emerge from competing or contradictory stakeholder interests. But in contrast to the warlike academy-industry binary that Hayhoe (2004) depicted, we believe that stakeholder theory can provide a more nuanced and intricate depiction of technical communication academic programs and how they interrelate with, rely on, and are even constituted by a broad network of stakeholders. Stakeholder theory offers an invaluable heuristic for identifying, evaluating, and reconciling the interests of an often complex array of forces that can claim a legitimate interest in our academic programs. In doing so, it productively works across the dichotomies of ideal versus material, theory versus practice, industry versus academy, and so on. Stakeholder theory can help us account for our programs' many constituent stakeholders and their relative salience, but it does not demand acquiescence to them. Perhaps most significantly, stakeholder theory implores us to remain aware of the impact our programs have upon their varied stakeholders and to avoid the separation fallacy, which stand as vitally important ethical injunctions.

To reiterate, in each of the examples we have provided, our point has not been to suggest a privileging of economic and material interests over all others. In fact, a stakeholder theoretical approach is necessarily agnostic about which stakeholder interests prevail, provided that the interests of all salient stakeholders are at least accounted for. We also concede that there nothing fundamentally new or revolutionary about stakeholder theory; it is just a different tool to describe the familiar mechanisms of rhetoric and politics that have always been operative. Berlin (2003), for instance, eloquently described a stakeholder approach to academic programs without once invoking the concept:

> A college curriculum is a device for encouraging the production of a certain kind of graduate, in effect, a certain kind of person. In directing what courses will be taken and in what order, a curriculum undertakes the creation of consciousness and behavior. A curriculum does not do this on its own, free of outside influence. Instead, it occupies a position between the conditions of the larger society it serves—the economic, political, and cultural sectors—and the work of teachers-scholars within the institution. . . . In short, the curriculum serves as a mediator between the demands of those outside the institution—employers, government agencies, political groups—and those within it—primarily faculty, the disciplines they serve, and students. The response of the curriculum to the exigencies of its historical moment thus represents a negotiation among forces both outside and inside the institution. (p. 18)

Stakeholder theory can serve as an important reminder that our institutions and academic programs are enmeshed in the complex political and material realities that Berlin describes. However, stakeholder theory additionally provides a framework for accounting for and engaging with such realities on a programmatic level. In doing so, it offers a broadly descriptive yet practical perspective that simpler models or reductive binaries alone cannot.

REFERENCES

Ackerman, F., & Eden, C. (2011). Strategic management of stakeholders: Theory and practice. *Long Range Planning, 44,* 179–196. doi:10.1016/j.lrp.2010.08.001

Alberts, D. J. (2007). Stakeholders or subject matter experts: Who should be consulted? *Energy Policy, 35,* 2336–2346.

Anderson, P. V. (1995). Evaluating academic technical communication programs: New stakeholders, diverse goals. *Technical Communication, 42,* 628–633.

Anderson, P. V. (2010). *Technical communication: A reader-centered approach.* Boston, MA: Wadsworth.

Bae, K.-H., Kang, J.-K., & Wang, J. (2011). Employee treatment and firm leverage: A test of the stakeholder theory of capital structure. *Journal of Financial Economics, 100,* 130–153. doi:10.1016/j.jfineco.2010.10.019

Barton, S. L., Hill, N. C., & Sundaram, S. (1989). An empirical test of stakeholder theory predictions of capital structure. *Financial Management, 18*(1), 36–44.

Beierle, T. C. (2002). The quality of stakeholder-based decisions. *Risk Analysis, 22,* 739–749.

Berlin, J. A. (2003). *Rhetorics, poetics, and cultures: Refiguring college English studies.* Lauer Series in Rhetoric and Composition. West Lafayette, IN: Parlor.

Billgren, C., & Holmén, H. (2008). Approaching reality: Comparing stakeholder analysis and cultural theory in the context of natural resource management. *Land Use Policy, 25,* 550–562. doi:10.1016/j.landusepol.2007.11.004

Bourne, L., & Walker, D. H. T. (2005). Visualising and mapping stakeholder influence. *Management Decision, 43,* 649–660.

Brugha, R., & Varvasovszky, Z. (2000). Stakeholder analysis: A review. *Health Policy and Planning, 15,* 239–246.

Bryson, J. M. (2003). What to do when stakeholders matter: A guide to stakeholder identification and analysis techniques. *Public Management Review, 6*(1), 21–53.

Campbell, B., & Mark, M. M. (2006). Toward more effective stakeholder dialogue: Applying theories of negotiation to policy and program evaluation. *Journal of Applied Social Psychology, 36,* 2834–2863. doi:10.1111/j.0021-9029.2006.00131.x

Carnegie, T. A. M. (2007). Integrating context into assessing US technical communication programs. *Technical Communication, 54,* 447–458.

Carter, M., Anson, C. M., & Miller, C. R. (2003). Assessing technical writing in institutional contexts: Using outcomes-based assessment for programmatic thinking. *Technical Communication Quarterly, 12,* 101–114.

Carroll, A. B. (1989). *Business and society: Ethics and stakeholder management.* Cincinnati, OH: Southwestern.

Clarkson, M. (1995). A stakeholder framework for analyzing and evaluating corporate social performance. *The Academy of Management Review, 20*(1), 92–117.

Conklin, J. (2008). Introduction to the special issue on qualitative research in technical communication. *Technical Communication, 55,* 329–332.

Coppola, N. (1997). A rhetorical analysis of stakeholders in environmental communication: A model. *Technical Communication Quarterly, 6,* 9–24.

Coppola, N. (2011). Professionalization of technical communication: Zeitgeist for our age introduction to this special issue (part 1). *Technical Communication, 58,* 277–284.

Coppola, N. (2012). Professionalization of technical communication: Zeitgeist for our age introduction to this special issue (part 2). *Technical Communication, 59,* 1–7.

Cragg, W. (2002). Business ethics and stakeholder theory. *Business Ethics Quarterly, 12,* 113–142.

Dayton, D. (2002). Evaluating environmental impact statements as communicative action. *Journal of Business and Technical Communication, 16,* 355–405. doi:10.1177/105065102236522

de Bakker, F. G. A., & de Hond, F. (2008). Introducing the politics of stakeholder influence. *Business & Society, 47,* 8–20. doi:10.1177/0007650307306637

de Bussy, N., Watson, R. T., Pitt, L. F., & Ewing, M. T. (2000). Stakeholder communication management on the Internet: An integrated matrix for the identification of opportunities. *Journal of Communication Management, 5,* 138–146. doi:10.1108/13632541011090419

Donaldson, T., & Preston, L. E. (1995). The stakeholder theory of the corporation: Concepts, evidence, and implications. *Academy of Management Review, 20,* 65–91.

Egels-Zandén, N., & Sandberg, J. (2010). Distinctions in descriptive and instrumental stakeholder theory: A challenge for empirical research. *Business Ethics: A European Review, 19,* 35–49. doi:10.1111/j.1467-8608.2009.01577.x

Eubanks, P., & Abbott, C. (2003). Using focus groups to supplement the assessment of technical communication texts, programs, and courses. *Technical Communication Quarterly, 12,* 25–45.

Fassin, Y. (2010). A dynamic perspective in Freeman's stakeholder model. *Journal of Business Ethics, 96,* 39–49. doi:10.1007/s10551-011-0942-6

Fraser, C., & Zhu, C. (2008). Stakeholder perception of construction site managers' effectiveness. *Construction Management and Economics, 26,* 579–590. doi:10.1080/01446190802036151

Freeman, R. E. (1984). *Strategic management: A stakeholder approach.* Marshfield, MA: Pitman.

Freeman, R. E. (1994). The politics of stakeholder theory: Some future directions. *Business Ethics Quarterly, 4,* 409–421.

Freeman, R. E. (2010). Managing for stakeholders: Trade-offs or value creation. *Journal of Business Ethics, 96,* 7–9. doi:10.1007/s10551-011-0935-5

Freeman, R. E., Harrison, J. S., Wicks, A. C., Parmar, B. L., & de Colle, S. (2010). *Stakeholder theory: The state of the art.* Cambridge, UK: Cambridge University Press.

Freeman, R. E., & Reed, D. (1983). Stockholders and stakeholders: A new perspective on corporate governance. *California Management Review, 15*(3), 88–106.

Gallagher, C. W. (2011). Being there: (Re)making the assessment scene. *College Composition and Communication, 62*(3), 450–476.

Grabill, J. T., & Simmons, W. M. (1998). Toward a critical rhetoric of risk communication: Producing citizens and the role of technical communicators. *Technical Communication Quarterly, 7,* 415–441.

Harner, S., & Rich, A. (2005). Trends in undergraduate curriculum in scientific and technical communication programs. *Technical Communication, 52,* 209–220.

Hartung, K. K. (1998). What are students being taught about the ethics of technical communication?: An analysis of the ethical discussions presented in four textbooks. *Journal of Technical Writing and Communication, 28,* 363–383.

Hayhoe, G. F. (2004). Inside out/outside in: Transcending the boundaries that divide the academy and industry. In T. Kynell-Hunt & G. J. Savage (Eds.), *Power and legitimacy in technical communication* (pp. 101–113). Amityville, NY: Baywood.

Ihlen, Ø., & Berntzen, Ø. (2007). When lobbying backfires: Balancing lobby efforts with insights from stakeholder theory. *Journal of Communication Management, 11,* 235–246.

Jennings, M. A. (1998). Teaching stakeholder theory: It's for strategy, not for business ethics. *The Journal of Legal Studies Education, 16,* 203–226.

Johnson-Eilola, J. (1996). Relocating the value of work: Technical communication in a post-industrial age. *Technical Communication Quarterly, 5,* 254–270.

Kline, J., & Barker, T. (2012). Negotiating professional consciousness in technical communication: A community of practice approach. *Technical Communication, 59,* 32–48.

Laplume, A. O., Sonpar, K., & Litz, R. A. (2008). Stakeholder theory: Reviewing a theory that moves us. *Journal of Management, 3,* 1152–1189. doi:10.1177/0149206308324322

Luoma-aho, V., & Paloviita, A. (2010). Actor-networking stakeholder theory for today's corporate communications. *Corporate Communications: An International Journal, 15,* 49–67.

Miller, C. (1989). What's practical about technical writing? In B. E. Fearing & W. K. Sparrow (Eds.), *Technical writing: Theory and practice* (pp. 14–24). New York, NY: Modern Language Association.

Miller, T. P. (1991). Treating professional writing as social praxis. *Journal of Advanced Composition, 11,* 57–72.

Mitchell, R., Agle, B. R. & Wood, D. I. (1997). Toward a theory of stakeholder identification and salience: Defining the principles of who and what really counts. *Academy of Management Review, 22,* 853–886.

Morgan, M. (1997). Evaluating technical communication programs: Collaborating for quality. In K. Staples & C. Ornatowski (Eds.), *Foundations for teaching technical communication: Theory, practice, and program design* (pp. 317–330). Greenwich, CT: Ablex.

Mygind, N. (2009). Stakeholder ownership and maximization. *Corporate Governance, 9,* 158–174.

Nugent, J. (2010). Certificate programs in technical communication: Through sophistic eyes. In D. Franke, A. Reid, & A. di Renzo (Eds.), *Design discourse: Composing and revising professional writing programs* (pp. 153–170). Anderson, SC: WAC Clearinghouse/ Parlor.

Orij, R. (2010). Corporate social disclosures in the context of national cultures. *Accounting, Auditing & Accountability, 23*(7), 868–889. doi: //10.1108/09513571011080162

Orts, E. W., & Strudler, A. (2002). The ethical and environmental limits of stakeholder theory. *Business Ethics Quarterly, 12,* 215–234.

Oxford English Dictionary. (2004). *Stakeholder*. Retrieved June 26, 2013, from http://dictionary.oed.com

Parekh, S. (2008). The crisis and emancipation of the modern corporate executive: How the Bhagavad Gita reinforces Edward Freeman's stakeholder theory. *The Michigan Journal of Business, 1*(1), 9–33.

Phillips, R. (2003). Stakeholder legitimacy. *Business Ethics Quarterly, 13,* 25–41.

Phillips, R., Freeman, R. E., & Wicks, A. (2003). What stakeholder theory is not. *Business Ethics Quarterly, 13,* 479–502.

Roberts, R. W., & Mahoney, L. (2004). Stakeholder concept of the corporation: Their meaning and influence in accounting research. *Business Ethics Quarterly, 14,* 399–431.

Rude, C. D. (2009). Mapping the research questions in technical communication. *Journal of Business and Technical Communication, 23,* 174–215. doi:10.1177/1050651908329562

Salvo, M. J., & Ren, J. (2007). Participatory assessment: Negotiating engagement in a technical communication program. *Technical Communication, 54,* 409–411.

Savage, G. J. (1996). Redefining the responsibilities of teachers and the social position of the technical communicator. *Technical Communication Quarterly, 5,* 309–327.

Schwartz, M. S. (2006). God as a managerial stakeholder? *Journal of Business Ethics, 66,* 291–306. doi:10.1007/s10551-005-5599-6

Scott, J. B. (2004). Rearticulating civic engagement through cultural studies and service-learning. *Technical Communication Quarterly, 3*(3), 289–306.

St.Amant, K. (2003). Expanding internships to enhance academic-industry relations: A perspective in stakeholder education. *Journal of Technical Writing and Communication, 33,* 231–241.

St.Amant, K., & Nahrwold, C. (2007). Acknowledging complexity: Rethinking program review and assessment in technical communication. *Technical Communication, 54*(4), 409–411.

Starik, M. (1995). Should trees have managerial standing? Toward stakeholder status for non-human nature. *Journal of Business Ethics, 14,* 207–217.

Wallace, C. et al. (2006). Student-based case studies in software communication. *Proceedings of the 19th conference on Software Engineering Education and Training Workshops.* doi:10.1109/CSEETW.2006.16

Wang, H. (2010). Factor analysis of corporate environmental responsibility: From the stakeholder theory perspective. *Environmental Development and Sustainability, 12,* 481–490. doi:10.1007/s10668-009-9206-3

Werhane, P. H. (2000). Business ethics, stakeholder theory, and the ethics of healthcare organizations. *Cambridge Quarterly of Healthcare Ethics, 9,* 169–181.

Wicks, A. C., Gilbert, D. R., & Freeman, R. E. (1994). A feminist reinterpretation of the stakeholder concept. *Business Ethics Quarterly, 4,* 475–497.

Wijnberg, N. M. (2000). Normative stakeholder theory and Aristotle: The link between ethics and politics. *Journal of Business Ethics, 25,* 329–342.

Yamak, S., & Süer, Ö. (2005). State as a stakeholder. *Corporate Governance, 5,* 111–120.

Zakhem, A. (2008). Stakeholder management capability: A discourse-theoretical approach. *Journal of Business Ethics, 79,* 395–405.

Zoetewey, M. W., & Staggers, J. (2004). Teaching the Air Midwest case: A stakeholder approach to deliberative technical rhetoric. *IEEE Transactions on Professional Communication, 47,* 233–243.

CHAPTER 2

A Technical Communication Venture in Building Academic-Entrepreneur Relations and Partnerships

John M. Spartz and Ryan P. Weber

ABSTRACT

This chapter argues that technical communication neglects the study and teaching of entrepreneurship by conceiving of academy-industry relationships too narrowly. Consequently, the writing tasks and rhetorical situations of entrepreneurs remain poorly understood and neglected in the classroom. We identify the entrepreneurship gap in technical communication and report the results of two research projects that begin to address this gap.

INTRODUCTION

When examining academy-industry partnerships, educators and researchers in technical communication typically focus on existing businesses and corporations. This perspective, however, is limiting. A full consideration of industry must include writing that entrepreneurs do to develop ventures that do not yet exist. We believe technical communication programs would benefit from paying more attention to entrepreneurship. It is a vibrant, multidisciplinary field with considerable resources and growing influence in and beyond the university.

Entrepreneurship programs and centers are emerging rapidly on American campuses to educate the increasing number of college students with entrepreneurial ambitions. According to statistics presented by the U.S. Small Business

Administration (2001), two thirds of college students intend to become entrepreneurs at some point in their careers, while an Intuit Future of Small Business Report (2007) argues that Generation Y "will emerge as the most entrepreneurial generation ever" (p. 5). As students pursue their entrepreneurial ambitions, they will encounter unique rhetorical situations and writing challenges that technical communication is well suited to address.

Our interest in entrepreneurship arose during the development and teaching of the class, *English 420E: Business Writing for Entrepreneurs*, offered by the Professional Writing Program at Purdue University. The curriculum was interesting and well received, but the projects, which involved pitching business ideas and writing business plans, were fairly standard to entrepreneurship courses across several disciplines. Thus, after teaching several iterations of the course, we suspected that we had failed to capitalize on the unique knowledge of technical communication. While considering alternative assignments, we discovered a lack of research about the rhetorical needs, habits, and practices of entrepreneurs. This situation creates difficulties for technical communication teachers who want to develop projects and courses targeting aspiring entrepreneurs. It also represents a significant research gap technical communication can and should address.

This chapter examines this gap via findings from projects in which we analyzed the technical communication and writing-related offerings of certain entrepreneurship programs and conducted interviews with eight entrepreneurs from a variety of industries. Working from these research results, we outline how technical communication might address the writing needs of entrepreneurs and suggest directions for future research.

PREVIOUS STUDIES ON ENTREPRENEURIAL WRITING AND COMMUNICATION

Few scholars discuss the relationships between entrepreneurship and technical communication, but existing studies provide direction for this project while demonstrating a significant gap in the technical communication scholarship.

Technical Communication Research on Entrepreneurial Writing

Certain scholars examine the writing conducted by entrepreneurs or entrepreneurial ventures. Doheny-Farina (1986, 1992), for example, has written about the role that texts play in shaping entrepreneurial ventures. His analysis (1992) of the successes and failures of a start-up business plan offers "a picture of a text that played a very important but not exclusive role in the rise and decline of a new organization" (p. 92). More recently, Mara (2008) analyzed the shifting ethos appeals that emerged in the marketing materials of a light jet start-up. Still others (e.g., Durack, 2003; Leydens, 2008; Windsor, 2006) examine entrepreneurship writing as part of larger historical or ethnographic projects involving workplace and technical writing.

Additionally, technical communication scholars have argued that entrepreneurs have unique communication needs worth addressing. Nunally and Saad (1993) explain that technopreneurs—high-tech professionals who leave established companies to start their own ventures—face unique communication challenges, as they are often undertrained in communication skills. Further, the autonomy of entrepreneurs often leaves them isolated from professional networks. Jablonski (2003), in turn, argues that technical communication classrooms focus too exclusively on preparing students for the rhetorical situations of Fortune 500 companies, yet many students find employment at smaller companies or through entrepreneurial ventures. This concern with preparing students for the writing tasks required by entrepreneurial endeavors matches our concern for providing rhetorical resources to this neglected population.

Other scholars find a relationship between the skills of technical communicators and the innovative spirit of entrepreneurship. Bekins and Williams (2006) argue that an entrepreneurial drive should manifest in technical communication pedagogy. Similarly, Faber and Johnson-Eilola (2002) claim technical communicators should envision themselves as innovative knowledge workers. Fisher, Bowers, Ellertson, Brumm, and Mickelson (2003) push these entrepreneurial implications further by arguing for "a more entrepreneurial role for the technical communicator and a more entrepreneurial feel for the technical-communication curriculum" (p. 314). Such studies provide ways to conceive the overlap between technical communication and entrepreneurship, but we argue that these connections have only begun to be forged.

We believe technical communication teachers and scholars have overlooked entrepreneurship for several reasons. For one, entrepreneurship is an emerging academic area, so it might have escaped the notice of technical communication scholars immersed in researching other types of writing. Additionally, the writing done by entrepreneurs often takes place within fledgling or prelaunch organizations; therefore, their work might not be as visible as writing coming from large, established organizations. Lastly, some scholars might shun this focus due to narrow definitions of entrepreneurship associated with conservative politics or cutthroat ambition. This chapter works with a more expansive definition of entrepreneurship that sees it as a holistic skillset that allows people to recognize opportunity, instigate change, and foster collaboration to create something new. Additionally, this chapter uses the word "entrepreneurship" to indicate any creation of a new business or technology versus scholars who prefer a narrower definition of the term to indicate ventures involving the commercialization of new goods, chains of supply, or markets (Schumpeter, 1934; Vesper, 1980).

Research on Writing for Business and Entrepreneurship

Many business researchers report that entrepreneurs need strong rhetorical, writing, and technical communication skills. Donckels (1991) and Sexton and Upton (1987) argue that entrepreneurship curricula should include more communication skills, and Lee and Jones (2008) found that entrepreneurs benefit from training in face-to-face and electronic communication. Other business scholars have examined the rhetorical strategies of entrepreneurs and note that technical communication skills could benefit aspiring entrepreneurs. Holt and McPherson (2010), for example, analyze the persuasive tactics of three entrepreneurs and find they use "a blend of logos, ethos and pathos in order to address a variety of constituents' interests simultaneously" (p. 22). Moreover, rhetoric helps create the institutional structures of an entrepreneurial venture.

In a study of 10 social-entrepreneurship firms, Ruebottom (2011) found that when organizations face resistance to proposed social-change programs, they generate support by creating narratives that frame their organizations as protagonists and their opposition as villains. The authors conclude that, because rhetorical strategy is so central to an organization's legitimacy, rhetoric should be "developed as a core competency" (p. 16) among social entrepreneurs. These conclusions are supported by studies analyzing the discursive resources entrepreneurs use to construct their identities (Watson, 2009), the importance of metaphor in the entrepreneurial process (Gaddefors, 2007; Hartelius & Browning, 2008), the value of communicating a company vision statement (Baum, Locke, & Kirkpatrick, 1998), and the persuasive tactics used during pitches to potential investors (Tsai & Lai, 2011). Taken together, this research provides insight into the rhetorical experiences and needs of entrepreneurs.

METHOD OF EXAMINATION

This chapter takes a 2-pronged approach to examine the communication needs and attitudes of entrepreneurs and the curricular focus of entrepreneurship-related technical communication pedagogy. The first phase of our study employed archival research to analyze academic programs and their writing-related offerings. The second phase involved interviews with practicing entrepreneurs and small-business owners.

Research on Programmatic Offerings

To better understand entrepreneurship education, we wanted to determine what writing/communication courses were offered for entrepreneurship students. We approached this phase of the project surmising that few entrepreneurship courses focus their writing instruction on anything beyond the well-known business plan. These courses lack a substantive investment in rhetoric or rhetorical skills as understood by technical communication and rhetorical studies. Few courses in technical communication specifically focus on the wide range of documents used by entrepreneurs. Thus, our first task was to determine the programmatic offerings of universities that support entrepreneurship programs.

As with other technical communication scholars interested in university curricular offerings (Harner & Rich, 2005; Nugent, 2010; Rainey, Turner, & Dayton, 2005), we surveyed the programmatic and departmental websites of 25 U.S. universities. To systematically select institutions, we used *Entrepreneur's* list of the "Top 25 Undergraduate Schools for Entrepreneurship Programs" provided by *The Princeton Review* (2011). Specifically, we accessed the websites for each program and searched those sites for information about entrepreneurship curriculum. Then we reviewed websites that listed courses in technical communication (primarily English and communication departments) to find classes that offered entrepreneurship-related course content. Thus, the accuracy of our analysis depends on the accuracy of those institutional websites.

When visiting entrepreneurship program sites, we accessed the course descriptions and other publicly accessible related information (i.e., syllabi) for all required courses. There, we searched for information about the type(s) of writing projects, instruction, or skills related to entrepreneurship. When accessing English and communication departmental websites in search of entrepreneur-related content, we investigated all courses, including those not designated as part of interdisciplinary entrepreneurship curricula or overtly labeled as entrepreneurship writing courses.

Qualitative Research Interviews

The role that research can play in building the relationship between the academy and industry is well established. According to Anderson, Brockmann, and Miller (1983), improved qualitative research into workplace writing has the potential for at least three positive outcomes for the field: building theory, supporting curriculum design, and informing the practices and problem-solving strategies of practicing professionals (p. 10). Doheny-Farina and Odell (1985) similarly argue that "qualitative studies of workplace writing contribute substantially to the understanding of the composing process, of the characteristics of good writing, of how readers make meaning, and of the functions of writing" (as cited in Sullivan & Spilka, 1992, 2010, p. 4). Halpern (1988), moreover, posits that qualitative research allows us to "discover meaning" (p. 30) and ask "new kinds of questions about how people communicate at work and also about what we teach in our classes," offering insight by "suggesting incongruities or gaps between what we thought we knew and what is really the case" (pp. 38–39). Furthermore, Sullivan and Spilka (1992) note that "qualitative research is the most accessible type of research to technical communicators" as it typically focuses on "describing how and why people act as they do in the workplace," which "has great potential to lead to practical applications that can benefit the field and assist in its continued growth" (Sullivan & Spilka, 1992, 2010, p. 5).

Following Steiner Kvale's (1996) discussion of qualitative research methodology, we set up interviews to help us interpret and describe central themes in the communication practices of our subjects. Drawing on this method, we focused on determining the types of documents functioning entrepreneurs produce, the types of communication skills that business founders consider important, and the writing tasks entrepreneurs perform to found, maintain, and grow their businesses. By conducting small-scale, one-on-one interviews with business founders, we hoped to unearth different documents and genres considered relevant by entrepreneurs.

Ultimately, we interviewed eight business owners in different industries: engineering, technology development, software, architecture, green-construction supplies, bike and ski sales, professional recruitment services, and restaurant management. Participants were asked 11 questions, including a general description of their business; reflection on the documents necessary to fund, start, maintain, and grow their businesses; the writing skills they value; the education/training they received to perform these tasks; and their writing-related advice to aspiring entrepreneurs.

ANALYSIS OF PROGRAMMATIC OFFERINGS

Our argument herein developed from our hypothesis that entrepreneurship programs and English departments similarly neglect entrepreneurial writing and communication. Thus, we designed our analysis to determine if entrepreneurship programs emphasize writing. From the available materials on campus websites, writing-centered curricula for entrepreneurs occur as isolated, rather than systemic, practices. Further, it appears that few English departments are forging connections with their entrepreneurship programs.

Ultimately, our research confirms that courses centered on business-plan writing are standard to entrepreneurship curricula. In this analysis, 20 of 25 programs reviewed listed a course dedicated to business plan creation (see Table 1), and it is highly likely that other programs also incorporate business plan writing in their curricula. Table 1 indicates the interdisciplinary nature of entrepreneurship programs, as the listed courses are housed in Entrepreneurship, Management, Management and Human Resources, Business, and Strategic Management departments.

Notably, English Departments rarely participate in these interdisciplinary programs, though the University of Utah's *WRTG 4001: Writing Business Plans and Proposals* course stands out: It is offered as part of a technical writing minor housed in the Writing program. This course represents the only connection between an Entrepreneurship program and an English/Technical Communication/Rhetoric department identified during this review process. One potential explanation for the existence of this "atypical" course is that *WRTG 4001* is offered only

Table 1. Entrepreneurship Courses in Business Plan Writing

University of Houston	ENTR 4360: Business Plan and Implementation
Baylor University	ENT 4325: Entrepreneurial Business Plan ENT 5334: Seminar in Business Plan Case Competition
Syracuse University	EEE 458: Business Plan Laboratory
University of Southern California	BAEP 454: The Business Plan
Northeastern University	ENTR 4501: Planning and Growth
BYU	MBA 590R: Business Plans
University of Arizona	ENTR 484: Development of New Venture Plans
University of Oklahoma	ENT 4603: New Venture Development
Temple University	SGM 3002: Business Plan Development
Drexel University	MGMT 365: Business Planning for Entrepreneurs
University of Missouri-Kansas City	ENT 327: Creating the New Venture: Experiential Learning
Miami University	ESP 467: New Venture Creation
University of Utah	WRTG 4001: Writing Business Plans and Proposals
DePaul University	MGT 370: Business Plan Development
University of Maryland, College Park	BMGT 465: Business Plan for the New Venture BUSI 758: Business Plan Writing and China Business Plan Competition Training
Belmont University	ETP 4500: Venture Planning
Texas Christian	MANA 40143: New Venture Development
University of Wisconsin, Madison	MHR 434: Venture Creation
Oklahoma State	EEE 3020: Business Plan Laboratory
Xavier	ENTR 311: New Venture Planning

in an online format. As such, it may be strategically positioned to attract working adults or potential small business owners in the greater Salt Lake City area. In all other cases, departments outside of English or Communication handle business-plan writing. This finding suggests technical communication teachers should offer writing courses that move beyond the business plan in order to maximize the value of their knowledge base.

An emphasis on writing and communication beyond the business plan emerged only sporadically among entrepreneurship curricula. Courses devoted to entrepreneurial communication are rare, with the exception of *MGT 9968: Entrepreneurial Communications—Selling and Negotiating* (City University of New York-Baruch campus). Based on the course description, the class appears to offer a more holistic approach to the rhetorical situations and audiences entrepreneurs encounter beyond the business plan. This rhetorical attention might be attributable to the course's status as an offering in the MBA in Entrepreneurship program at CUNY-Barauch.

Other courses emphasize communication skills, such as Northwestern University's *TECE 6340: The Technical Entrepreneur as Leader and Innovator*; its catalog description includes a line explaining that the course "stresses communications skills, both written and oral, along with self-discovery." A few other courses appear to offer a rhetorical approach, even if writing and communication are not directly referenced in their descriptions. For instance, Belmont University in Nashville, Tennessee, offers *ETP 3600: Building Stakeholder Relationships*, which may focus on persuasion, communication networks, documents, and rhetorical action related to obtaining stakeholder involvement.

Aside from these few classes, an emphasis on writing seems neglected or scattered among various courses. Many programs involve courses in marketing, which might include the creation of marketing plans and advertisements; human resources, which may involve the authoring of job descriptions or policy manuals; and law, which may cover policy or patent writing.

Overall, our research indicates that business-plan writing is extensively covered by entrepreneurship programs, but the broader rhetorical needs of entrepreneurs are not thoroughly addressed. Currently, writing and rhetorical skills are neglected in many programs surveyed here, and important genres remain overlooked. Many entrepreneurship writing programs do not fully incorporate technical communication skills in their curricula, and few English/technical communication programs directly address the needs of our target population. These findings suggest that technical communication has an unrealized opportunity to provide expertise within entrepreneurship programs.

INTERVIEW FINDINGS

The entrepreneurs we interviewed identified writing needs that emerged as they created and maintained their businesses. Although differences existed in the level of writing they performed and the rhetorical situations they faced, the majority of our interviewees valued writing as part of their business creation and growth.

Entrepreneur Writing Needs and Habits

In the interviews, significant patterns emerged that might help technical and professional communicators identify the writing *needs* and *habits* of entrepreneurs. On the whole, the entrepreneurs we interviewed

- Write many documents themselves;
- Value good, effective writing;
- Express insecurity about their writing abilities;
- Desire writing training customized for entrepreneurs;
- Use various documents and genres in their businesses.

When describing writing *traits* they valued, subjects repeatedly emphasized correctness (reflecting the credibility of the venture) and conciseness (which presents rhetorical and legal challenges).

Often, entrepreneurs adapt preexisting documents to new rhetorical situations instead of writing each document from scratch. In producing documents, entrepreneurs spend considerable time gathering and synthesizing information about their industry, their market, their community, and their competition. Many interviewees also reported creating strategic narratives about themselves and their businesses, which echoes research on the importance of narrative in entrepreneurial ventures.

Write Many Documents

Most of the entrepreneurs interviewed reported spending ample time writing documents and that writing is a necessity for their work. For example, an engineer running an engineering consulting firm stated that writing occupied a significant percentage of his time, noting that "in engineering, probably like a lot of businesses, you do a whole lot more writing than you would have thought when you were in college. I would think 20–25% of your time could be writing letters." Though his firm employs 60 individuals, he still devotes a large portion of his time to writing support documents for project proposals. Further, an architect who runs his own small firm likewise reported that he spends a substantial amount of time "writing, writing, writing, everything you can to make it look like you know what you're doing."

Several interviewees reported that their writing demands escalate when they expand into new markets or develop new products or services. When reflecting on a previous venture, for example, an IT entrepreneur reported that "in my last business, we started a significant services organization. A tremendous amount of writing related to this was required." The Assistant Director of SCORE Minnesota, who has started and successfully run three businesses, argued that the desire to create coordinated social media campaigns using blogs and Twitter significantly increases the writing demands on entrepreneurs who want to use these tools to connect with customers and draw them to the company website.

Value Good Writing

Those entrepreneurs who frequently write also cited its importance for their venture's success. Most of these entrepreneurs consider strong writing skills a necessity and made direct correlations between writing ability and financial success. The founder of a professional recruiting firm even listed it as the most important skill in business, reporting that "writing and communication is number one for business, to be successful . . . what you write is so important. You have to be very careful about what you write."

Several subjects mentioned that writing is important because it represents the company, making or breaking its reputation. They seemed especially aware that poor writing can affect a company's reputation. The founder of the engineering firm expressed a commonly held sentiment, stating "anything we send out of here is a reflection on our company, so we try to make sure it's well written." Other entrepreneurs noted the importance of effective language in helping companies perform tasks and generate sales. The IT entrepreneur commented on the importance of quality technical procedures, contracts, and proposals by arguing "the thoroughness and clarity of these documents directly affected the sales results."

Express Writing Insecurity

Most interviewees did not speak highly of their writing skills. Their comments indicate awareness that writing is not their strength and that their training, skills, and experience do not meet the rhetorical demands created by their ventures. The IT professional maintained that he struggled to translate ideas into prose, reflecting,

> I had very little formal writing education, and one of the things I struggled with most was the process of writing. Most entrepreneurs I know struggle with the same thing. They are very passionate when talking about the business but struggle with actually getting it down on paper.

The owner of the architecture firm also indicated that his insecurity about writing caused indecision over how much information to provide potential customers. He also found himself unprepared to assess whether his documents were having the desired rhetorical effect among his audiences.

Some subjects hired employees with strong communication skills to address their perceived inadequacies. The SCORE Director explained, "I have to look for good people with those writing skills, because I don't have good enough skills—I don't think—to communicate those messages. I didn't get enough of that in college myself." Several interviewees wished that they had taken more writing classes in college, while the owner of a green home-goods store relied heavily on the skills developed in a college business-writing course.

Desire More Education and Training

Some interviewees had taken entrepreneurship classes and clinics offered by universities, SCORE, small business development centers, or other organizations. These classes often focus on writing a business plan, and many subjects expressed interest in an entrepreneurship-specific writing course. As the recruiting professional we interviewed explained,

> I think a writing class for entrepreneurs would be awesome. You know, in that class, for people like myself, a one-stop shop would be great. Writing the business plan, 'cause that's business-speak; writing the marketing copy . . . selling and promotional, strategic planning, social media. I have no idea how to write for social media, a blog or twitter or something, but that's all stuff that people need to do. PR, all that stuff. If you could have it in one class, that would be awesome.

The recruiting professional's proposed course content identifies documents entrepreneurs produce, and many other genres were noted during the interviews.

Use Various Documents and Genres

In starting, maintaining, and growing their businesses, these entrepreneurs relied on different texts like business plans, patents, cash-flow statements, proposals, letters, press releases, contracts, grant applications, operating procedures, and advertisements. Many of these documents describe the business to investors, vendors, potential customers, and even the company's own employees. The entrepreneurs in our research value the ability to successfully describe their businesses to multiple audiences. For example, the owner of a green home-goods store reported that she spends much of her time writing concise descriptions that convey the ecofriendly nature of her business to prospective vendors. The architect noted that he sends out single-sheets and substantial packets displaying the capabilities of his firm. The IT entrepreneur asserted that internal organizational documents, while often neglected, provide necessary cohesion and vision for the investors, customers, and employees of a startup:

> Standard employee guidebooks that contain not only policy and procedures, but company core values, mission statements, etc. are vital when you are

> growing a company. Often dismissed as requirements for Fortune 500 business, I believe these are even more important in a startup business, as you are typically moving very fast and cannot afford to have anyone that doesn't have clarity about who we are, what we are doing, and how we are performing against plan.

Writing Traits: Correctness and Conciseness

Interviewees repeatedly emphasized two writing traits: correctness and conciseness. However, nuances in their discussion indicate that these entrepreneurs have sophisticated conceptions of both traits and their relationship to specific rhetorical situations. For instance, almost all the entrepreneurs emphasized the importance of correct grammar and spelling. Interestingly, their discussion indicates that they are not concerned about correctness for its own sake; rather they care about the credibility factors associated with good grammar. The green home-goods storeowner summarized this sentiment best by noting that documents containing spelling, punctuation, or grammar errors would "kill their reputation right away."

Entrepreneurs revealed a nuanced understanding of conciseness and acknowledged that conciseness requires difficult decisions about what to include or omit. Many of these entrepreneurs must balance between including enough information to fully describe their businesses and keeping documents short enough to garner attention. The engineer encapsulated this dilemma, observing that "you want [your document] to be to the point so someone would read it, but yet you wanted to highlight what you thought were the important things." The architect echoed the pressure of being concise and comprehensive. However, he further emphasized that conciseness presents a liability challenge by limiting the inclusion of legally relevant information:

> There's so much information overload that people don't have time to read all this stuff. They want things summarized. The danger of that is that summarizing something, if you don't cover yourself, in my position in my business if you don't have the stuff on paper, you could end up getting in a lawsuit and end up not having enough information to cover yourself.

Other entrepreneurs write concisely because they understand audiences read lengthy documents for key information and ignore sections not relevant to their needs. Many considered the reading habits and information needs of audiences. The SCORE Director noted that he had only a few paragraphs for a business plan to connect with important audiences:

> And the first thing those bankers simply say is "I get a 1,000 of these things a month" and they're going to read the executive summary . . . it's very critical. If the executive summary doesn't connect to them, they're not going to read the rest of the document.

The same professional also had a strong sense of which business plan sections (the cash flow, the balance sheet, the profit and loss statement) investors would examine first to determine the viability of the business. The owner of the green home-goods store reported that she presents only the information readers would skim for: "My belief is less information, less filler, the better," arguing that readers are "going to look for the highlights, so my philosophy is give them the highlights."

Document Customization

These entrepreneurs also showed a preference for customizing existing documents for new audiences rather than creating documents from scratch. Several interviewees kept a stockpile of past documents that could be modified for new rhetorical situations. The green home-goods storeowner, for instance, mentioned the importance of keeping past documents as references for new writing demands. This strategy suggests that entrepreneurs try to save time while remaining rhetorically savvy. They rely not on inflexible templates, but on chunks of recurring text repurposed for new contexts. The engineer estimated that most of his company's proposals started from the same body before receiving customization: "We always try to tailor it to the specific request, even though 80% of it may be somewhat boilerplate. We may just have to rewrite it or put it in a different format." The IT entrepreneur mentioned a similar strategy as an effective way to compose customizable business plans:

> The business plan documents I wrote in the early days were very generic. When I was looking for funding, I was meeting with a wide variety of potential investors: institutional VC's, angels, regional funds etc. Each of these audiences really evaluates the business differently. A better approach would be to segment the plan into very brief and interchangeable sections that could be customized based on the audience.

Required Research

When creating texts, entrepreneurs see themselves as gathering and synthesizing a tremendous amount of information. Because business ideas and plans are speculative, entrepreneurs often have only their own credibility and past research as a foundation for new ideas. Therefore, many interview subjects emphasized the importance of entrepreneurs doing their homework before meeting with investors. The SCORE Director argued that when entrepreneurs meet with funders,

> they're going have to be able to communicate the research that they've done and the results of that research back to these organizations so they can show some validation that there is a market for what they want to sell and that they've thought through the process.

The IT entrepreneur made a similar point: "Particularity in the technology sector, the business models can be complex and sometimes hard to believe. Credible sources of information can be key. Referencing independent research is important." These observations match Chen, Yao, and Kotha's (2009) findings that an entrepreneur's level of preparedness for funding presentations positively correlates with venture capitalist funding decisions.

Success Narratives

Several entrepreneurs expressed interest in creating success narratives about themselves and their businesses. Sometimes these narratives involve the origin and journey of the business. The recruiting professional explained that

> they want to know "what's the story behind it?" Especially if you want to get PR and things like that, you've got to have a story, like "how did you get in the business?" Everybody wants to know that. You've got to create a compelling story.

Other times, creating a narrative involves crafting a story that attracts involvement, as "translating the success or failure of an endeavor is important. The communication of a success story is one of the most important things."

In discussing narratives, these entrepreneurs reflect scholarship (Down & Warren, 2006; Downing, 2005; Gartner, 2010; Johansson, 2004; Jones, Latham, & Betta, 2008; Rae, 2004; Steyaert, 1997; Steyaert & Bouwen, 1997) that explores the narrative strategies of entrepreneurs and the conceptualization of entrepreneurship as a narrative. For instance, research by Martens, Jennings, and Jennings (2007) demonstrates that "storytelling is a key mechanism through which entrepreneurs can leverage their existing capital to acquire additional resources" (p. 1125). This conclusion is echoed by academic researchers and by our interview subjects.

DISCUSSION

These findings demonstrate that many entrepreneurs value good writing. However, our analysis of programmatic offerings indicates that entrepreneurship curricula have not incorporated these practitioner attitudes toward writing. While this disconnect merits further research, it might result from the demands on entrepreneurship curriculum to cover copious amounts of material and a lack of research about the writing needs of entrepreneurs. Regardless, this gap offers technical communication an opportunity to build interdisciplinary and community partnerships. Our interview research begins to provide a window into how technical communication theory and teaching might address the writing and rhetorical needs of aspiring and practicing entrepreneurs.

Within this context, our research identified skills and concepts that might be incorporated in a technical communication class for entrepreneurs. These skills and concepts complement the strengths of successful technical writing curricula. However, we argue that technical writing curricula could focus specifically on entrepreneurs by emphasizing rhetorically savvy document customization, business and technology description, nuanced writing characteristics, reputation construction and management, and narrative construction.

Rhetorically Savvy Document Customization

Several interviewees indicated that a key component of their writing process is customizing audience-specific documents from previously written texts. While this process saves time and effort, it can lead to ineffective documents. Technical communication can offer students audience-analysis approaches that encourage savvier composing. With its recent interest in single-sourcing, technical communication scholarship has begun to develop sophisticated approaches to the challenging task of repurposing content.

Albers (2003) provides a model for analyzing the audiences of custom documents, and Swarts (2010) argues for a rhetorical approach to repurposing content. A curriculum emphasizing document customization also fits Honig's (2004) model of contingency-based business planning for entrepreneurial education. A technical communication pedagogy offering these and similar skills could effectively serve entrepreneur needs.

Business and Technology Descriptions

Our findings further indicate that entrepreneurs spend tremendous amounts of time describing their businesses, products, services, and processes to investors, customers, and employees. Such descriptions can be particularly challenging when the venture involves new or complex technology. However, writing business and technical descriptions is a mainstay of technical communication curricula, as technical communication instructors and practitioners often have much expertise in this arena. Teaching nascent entrepreneurs how to describe their businesses and technologies for various audiences could serve as a very productive exercise.

Nuanced Writing Characteristics

Our interviewees expressed concern about grammatical and mechanical correctness in writing. This focus, however, reflects a concern for company reputation rather than a desire for grammatical purity. Interviewees expressed anxiety about the impact of their errors, and technical communication likewise considers error from a rhetorical perspective. Technical communication also offers the approach advocated by Beason (2001) in his study of business professionals' reaction to errors:

> I do not believe students can understand error unless they and teachers alike better comprehend error in terms of its impact—not just textual conventions defining errors, not just categories or rankings of errors, but the ways in which errors manage to bother nonacademic readers. We must understand the experiences of readers as they encounter error, for errors are created in the mind as much as in the text (p. 34).

Thus, technical communication is well positioned to help entrepreneurs address grammatical concerns in effective ways and from a rhetorical perspective.

Reputation Construction and Management

Business owners display tremendous interest in developing and maintaining the reputation and identity of their venture. With its long-standing interest in ethos, technical communication can inform aspiring entrepreneurs about the complexities of corporate reputation. Calls for the use of ethos in technical communication pedagogy (Campbell, 1995; Newman, 1999; Reynolds, 1992; Spartz, 2010) and investigations of the construction of ethos in professional contexts and documents (Beason, 1991; Isaksson, 2010; Mara, 2008; Payne, 2000; Spoel, 2008) provide technical communication with a strong understanding of the factors influencing reputation. This approach also allows technical communicators to engage in analyses of ethos along several rhetorical dimensions (e.g., the practical wisdom, virtue, and goodwill of a corporation). Technical communication can also teach aspiring entrepreneurs how to use rhetorical concepts to conduct research and present data in a way that bolsters credibility.

Narrative Construction

Our interview findings indicate that entrepreneurs rely on carefully crafted narratives to build strong customer, investor, and community relations. Technical communication can, in turn, provide entrepreneurs with a well-developed and sophisticated understanding of narrative construction and analysis. Several technical communication scholars (Perkins & Blyler, 1997; Rentz, 1992) argue for the importance of narrative to professional writing. Others have analyzed the narratives of professional texts (Nilsson, 1997; Smart, 1999; Spoel, Goforth, Cheu, & Pearson, 2009) in a manner that could provide entrepreneurs with a background and methodology for helping students analyze and construct powerful and ethical entrepreneurial success stories. O'Connor's (2002) ethnographic research, for example, provides a taxonomy that might offer an important view into the creation and use of entrepreneurial narrative.

IMPLICATIONS AND CONCLUSION

Our experiences researching and collaborating across technical communication and entrepreneurship indicate that several opportunities for partnerships exist.

Academic Implications

Exposure and research into entrepreneurial practice is advocated as a way to narrow the "academia vs. business incongruence" (Kuratko, 2005, p. 586). It also allows us to shift the emphasis from educating "about" entrepreneurship to educating "for it" (Kirby, 2004), especially insofar as writing instruction is concerned (as cited in Edelman, Manolova, & Brush, 2008, p. 57). Our investigation suggests that technical communication can align its pedagogy with real-world entrepreneurial writing and communication practices, as technical communication can contribute unique knowledge, skills, and methods provided by the discipline. Further, this research provides technical communication teachers and various entrepreneurship stakeholders with the information needed to develop writing curricula aligned with practitioner values, needs, and desires.

As discussed, the traditional pedagogical focus on assigning the business plan as the only substantive piece of entrepreneur-related writing requires reevaluation. This reconsideration is salient, considering "neither the teaching of business plans, nor the plans themselves, are sufficiently justified on the basis of theoretical or empirical literature" (Honig, 2004, p. 258). Our research establishes a gap in institutional entrepreneurship offerings that technical communication can fill. It also identifies the real communication needs and values of entrepreneurs. This type of research can thus help bridge academy-industry relationships while facilitating cross-institutional connections between technical communication (e.g., English Departments) and programs in entrepreneurship housed in the business school or a freestanding center.

Once established, this academic connection provides technical communication educators and their departments with a variety of new opportunities. Specifically, we have found that articulating and supporting our value to entrepreneurship education better serves the needs of our nascent entrepreneur students (by inculcating a germane and transferable skillset). It also better serves the needs of technical communication itself.

Programmatic Implications

Technical communication might also benefit from leveraging these relationships into programmatic development, expansion, and maintenance. By building and nurturing relationships with departments outside of our English or communication homes, technical communication stands to gain a steady influx of entrepreneurship students taking courses in our department. Additionally, attracting students from non-English/technical communication majors, disciplines, and concentrations will facilitate our ability to teach the types of courses that best capitalize on our areas of interest and research. Because students interested in entrepreneurship major in a variety of fields, courses of this nature will be attractive to students across the university.

While not widespread, this type of interdisciplinary curricular development and student recruitment has proven successful. For example, in the fall of 2007, the Professional Writing Program at Purdue University began a relationship with the Burton D. Morgan Center for Entrepreneurship and Innovation and offered an elective course (*ENGL 420E: Business Writing for Entrepreneurs*) as part of the cross-disciplinary Certificate in Entrepreneurship curriculum. Establishing three sections of 420E per semester, the English Department and Professional Writing Program increased the number of students it served per academic year. This increase has implications for technical communication scholars, departments, and related academic units.

Further, in programs that offer a minor or certificate in technical communication (let alone, a major), a set of technical-communication-for-entrepreneurs courses potentially serves two purposes. First, these courses provide more breadth in offerings to students. Second, these courses stand to attract more students to our minor or certificate programs. After taking our courses, these students might discover that technical communication helps prepare them for success in the business world and continue to enroll in our classes.

Academic Contributions to Industry: Building a Relationship

The next phase of the research reported here is to generate a comprehensive survey to quantitatively gauge the landscape of entrepreneurial communication and writing. At present, we have developed relationships with our local SBDCs, business incubators, and entrepreneur groups. Using contact lists from these partners, the Directors of our regional entrepreneur groups will disseminate a survey to entrepreneurs and an e-mail requesting assistance with this research endeavor.

Our ability to develop affiliations with these Directors, and their willingness to assist in this type of research, illustrates how mutually beneficial work like ours can be in helping to build relationships across and beyond campus. The data that we generate, for example, can help these entrepreneur groups develop client services for the nascent and practicing entrepreneurs and small business owners in the region. This data-sharing represents one form of "exchanging expertise," a two-way model of influence whereby the research performed by academics shapes industry practices while industry practices shape academic program development (Rehling, 2004).

In an interview with the Director of Business Development at a regional business incubator, we learned of the industry value that our research might have. In our conversation, the Director discussed that the organization was in the midst of developing a technical training program as part of their national certification for business incubation management. Part of that program's focus is on providing client services to members about the writing and communication needs of stage-one and stage-two companies, differentiating the education based on those stages. Our survey, which will ask preliminary questions about business type, number of employees, and company revenue, will allow us to analyze and report results according to business stage. These results can have an immediate impact on the curricula of these entrepreneur training programs.

On the whole, the academy-industry implications discussed here only begin to delineate how technical communication can contribute to entrepreneurship education and curricula. Further, we have just begun to unearth the potential and mutual benefits that research into entrepreneurial communication might have. Our continued research and resulting rapport-building holds much potential for technical communication and the students, communities, and industry stakeholders with whom we interact. Our future work will certainly further illuminate the specific rhetorical and writing needs of entrepreneurs; however, our current research offers a first step in defining how technical communication can create relevant and unique entrepreneurship curriculum coursework and create effective academy-industry relationships.

REFERENCES

Albers, M. (2003). Multidimensional audience analysis for dynamic information. *Journal of Technical Writing and Communication, 33*(3), 263–279.

Anderson, P. V., Brockmann, J., & Miller, C. R. (1983). Introduction. In P. V. Anderson, J. Brockmann, & C. R. Miller (Eds.), *New essays in technical and scientific communication: Research, theory, practice* (pp. 7–14). Amityville, NY: Baywood.

Baum, J., Locke, E., & Kirkpatrick, S. (1998). A longitudinal study of the relation of vision and vision communication to venture growth in entrepreneurial firms. *Journal of Applied Psychology, 83*(1), 43–54.

Beason, L. (1991). Strategies for establishing an effective persona: Ethos in business speeches. *Journal of Business Communication, 28*(4), 326–346.

Beason, L. (2001). Ethos and error: How business people react to errors. *College Composition and Communication, 53*(1), 33–64.

Bekins, L., & Williams, S. (2006). Positioning technical communication for the creative economy. *Technical Communication, 53*(3), 287–295.

Campbell, C. P. (1995). Ethos: Character and ethics in technical writing. *IEEE Transactions on Professional Communication, 38*(3), 132–138.

Chen, X., Yao, X., & Kotha, S. (2009) Entrepreneur passion and preparedness in business plan presentations: A persuasion analysis of venture capitalists' funding decisions. *Academy of Management Journal, 52,* 199–214.

Doheny-Farina, S. (1986). Writing in an emerging organization: An ethnographic study. *Written Communication, 3*(2), 158–185.

Doheny-Farina, S. (1992). *Rhetoric, innovation, technology: Case studies of technical communication in technology transfer*. Cambridge, MA: MIT Press.

Doheny-Farina, S., & Odell, L. (1985). Ethnographic research on writing: Assumptions and methodology. In L. Odell & D. Goswami (Eds.), *Writing in nonacademic settings* (pp. 503–535). New York, NY: Guilford.

Donckels, R. (1991). Education and entrepreneurship experiences from secondary and university education in Belgium. *Journal of Small Business and Entrepreneurship, 9*(1), 35–42.

Down, S., & Warren, L. (2008). Constructing narratives of enterprise: Cliches and entrepreneurial self-identity. *International Journal of Entrepreneurial Behaviour & Research, 14*(1), 4–23.

Downing, S. (2005). The social construction of entrepreneurship: Narrative and dramatic processes in the coproduction of organizations and identities. *Entrepreneurship Theory and Practice, 29,* 185–204.

Durack, K. (2003). Observations on entrepreneurship, instructional texts, and personal interaction. *Journal of Technical Writing and Communication, 33*(2), 87–109.

Edelman, L. F., Manolova, T. S., & Brush, C. (2008). Entrepreneurship education: Correspondence between practices of nascent entrepreneurs and textbook prescriptions for success. *Academy of Management Learning and Education, 7*(1), 56–70.

Faber, B., & Johnson-Eilola, J. (2002). Migrations: Strategic thinking about the future(s) of technical communication. In B. Mirel & R. Spilka (Eds.), *Reshaping technical communication: New directions and challenges for the 21st century* (pp 135–148). Mahwah, NJ: Lawrence Erlbaum.

Fisher, D., Bowers, T., Ellertson, A., Brumm, T., & Mickelson, S. (2003). As the case may be: The potential of electronic cases for interdisciplinary communication instruction. *IEEE Transactions on Professional Communication, 46*(4), 313–319.

Gaddefors, J. (2007). Metaphor use in the entrepreneurial process. *International Journal of Entrepreneurial Behaviour & Research, 13*(3), 173–193.

Gartner, W. (2010). A new path to the waterfall: A narrative on a use of entrepreneurial narrative. *International Small Business Journal, 28*(1), 6–19.

Halpern, J. (1988). Getting in deep: Using qualitative research in business and technical communication. *Journal of Business and Technical Communication, 2*(2), 22–43.

Harner, S., & Rich, A. (2005). Trends in undergraduate curriculum in scientific and technical communication programs. *Technical Communication, 52*(2), 209–220.

Hartelius, E. J., & Browning, L. (2008). The application of rhetorical theory in managerial research: A literature review. *Management Communication Quarterly, 22*(1), 13–39.

Holt, R., & Macpherson, A. (2010). Sensemaking, rhetoric and the socially competent entrepreneur. *International Small Business Journal, 28*(1), 20–42.

Honig, B. (2004). Entrepreneurship education: Toward a model of contingency-based business planning. *Academy of Management Learning and Education, 3*(3), 258–273.

Intuit. (2007). Future of small business report. Retrieved October 15, 2012, from http://www.intuit.com/

Isaksson, M. (2010). Communicating corporate ethos on the Web: The self-presentation of PR agencies. *Journal of Business Communication, 47*(2), 119–140.

Jablonski, J. (2003). [Review of the book The new careers: Individual action and economic change, by M. B. Arthur, K. Inkson, & J. K. Pringle.] *Technical Communication Quarterly, 12*(2), 230–234.

Johansson, A. (2004). Narrating the entrepreneur. *International Small Business Journal, 22*(3), 273–293.

Jones, R., Latham, J., & Betta, M. (2008). Narrative construction of the social entrepreneurial identity. *International Journal of Entrepreneurial Behaviour & Research, 14*(5), 330–345.

Kirby, D. A. (2004). Entrepreneurship education: Can business schools meet the challenge? *Education and Training, 46*(8/9), 510–519.

Kuratko, D. F. (2005). The emergence of entrepreneurship education: Development, trends, and challenges. *Entrepreneurship Theory and Practice, 29*(5), 577–598.

Kvale, S. (1996). *InterViews: An introduction to qualitative research interviewing.* Thousand Oaks, CA: Sage.

Lee, R., & Jones, O. (2008). Networks, communication and learning during business start-up: The creation of cognitive social capital. *International Small Business Journal, 26*(5), 559–594.

Leydens, J. (2008). Novice and insider perspectives on academic and workplace writing: Toward a continuum of rhetorical awareness. *IEEE Transactions on Professional Communication, 51*(3), 242–263.

Mara, A. (2008). Ethos as market maker: The creative role of technical marketing communication in an aviation start-up. *Journal of Business and Technical Communication, 22*(4), 429–453.

Martens, M. L., Jennings, J. E., & Jennings, P. D. (2007). Do the stories they tell get them the money they need?: The role of entrepreneurial narratives in resource acquisition. *Academy of Management Journal, 50,* 1107–1132.

Newman, S. (1999). Aristotelian rhetorical theory as a framework for teaching scientific and technical communication. *Journal of Technical Writing and Communication, 29*(4), 325–334.

Nilsson, S. (1997). Understandability of narratives in annual reports. *Journal of Technical Writing and Communication, 27*(4), 361–375.

Nugent, J. (2010). Certificate programs in technical writing: Through sophistic eyes. In D. Franke, A. Reid, & A. DiRenzo (Eds.), *Design discourse: Composing and revising programs in professional and technical writing* (pp. 153–170). Anderson, SC: Parlor.

Nunally, P., & Saad, S. (1993). *Technical and business communication for entrepreneurs.* 1993 International Professional Communication Conference—The new face of technical communication: People, processes, products. Philadelphia, PA: IEEE Professional Communication Society.

O'Connor, E. (2002). Storied business: Typology, intertextuality, and traffic in entrepreneurial narrative. *The Journal of Business Communication, 39*(1), 36–54.

Payne, K. (2000). Speaking ebonics in a professional context: The role of ethos/source credibility and perceived sociability of the speaker. *Journal of Technical Writing and Communication, 30*(4), 367–384.

Perkins, J. M., & Blyler, N. R. (Eds.). (1999). *Narrative and professional communication* (Vol. 10). Portsmouth, NH: Greenwood.

Princeton Review. (2011). *Top entrepreneurship programs.* Retrieved October 20, 2012, from www.princetonreview.com/entrepreneur/

Rae, D. (2004). Practical theories from entrepreneurs' stories: Discursive approaches to entrepreneurial learning. *Journal of Small Business and Enterprise Development, 11,* 195–202.

Rainey, K., Turner, R., & Dayton, D. (2005). Do curricula correspond to managerial expectations?: Core competencies for technical communicators. *Technical Communication, 52*(3), 323–352.

Rehling, R. (2004). Reconfiguring the professor-practitioner relationship. In T. Kynell-Hunt & G. Savage (Eds.), *Power and legitimacy in technical communication: Volume II Strategies for professional status* (pp 87–100). Amityville, NY: Baywood.

Rentz, K. (1992). The value of narrative in business writing. *Journal of Business and Technical Communication, 6,* 293–315.

Reynolds, J. (1992). Classical rhetoric and the teaching of technical writing. *Technical Communication Quarterly, 1*(2), 63–76.

Ruebottom, T. (2011). The microstructures of rhetorical strategy in social entrepreneurship: Building legitimacy through heroes and villains. *Journal of Business Venturing, 26,* 1–19.

Schumpeter, J. (1934). *The theory of economic development.* New York, NY: Oxford University Press.

Sexton, D., & Upton, N. (1987). Evaluation of an innovative approach to teaching entrepreneurship. *Journal of Small Business Management, 25*(1), 35–43.

Smart, G. (1999). Storytelling in a central bank: The role of narrative in the creation and use of specialized economic knowledge. *Journal of Business and Technical Communication, 13*(3), 249–273.

Spartz, J. M. (2010, May). Rhetorical savvy as social skill: Modeling entrepreneur identity construction within educational content management systems. *The Writing Instructor*. Retrieved October 20, 2012, from http://www.writinginstructor.com/spartz

Spoel, P. (2008). Communicating values, valuing community through health-care websites: Midwifery's online ethos and public communication in Ontario. *Technical Communication Quarterly, 17*(3), 264–288.

Spoel, P., Goforth, D., Cheu., & Pearson, D. (2009). Public communication of climate change science: Engaging citizens through apocalyptic narrative explanation. *Technical Communication Quarterly, 18*(1), 49–81.

Steyaert, C. (1997). A qualitative methodology for process studies of entrepreneurship: Creating local knowledge through stories. *International Studies of Management and Organizations, 27*(3), 13–33.

Steyaert, C., & Bouwen, R. (1997). Telling stories of entrepreneurship: Towards a narrative-contextual epistemology for entrepreneurial studies. In R. Donckels & A. Miettinen (Ed.), *Entrepreneurship and SME research: On its way to the next millennium* (pp. 47–62). Aldershot, UK: Ashgate.

Sullivan, P., & Spilka, R. (1992). Qualitative research in technical communication: Issues of value, identity, and use. *Technical Communication, 39*(4), 592–606.

Sullivan, P., & Spilka, R. (2010). Qualitative research in technical communication: Issues of value, identity, and use. In J. Conklin & G. F. Hayhoe (Eds.), *Qualitative research in technical communication* (pp. 1–24). New York, NY: Routledge. (Reprinted from *Technical Communication, 39*(4), 592–606, 1992.)

Swarts, J. (2010). Recycled writing: Assembling actor networks from reusable content. *Journal of Business and Technical Communication, 24,* 127–163.

Tsai, W., & Lai, Y. (2011). Persuasion methods of fund raising for new ventures. *Technology management in the energy smart world (PICMET), 2011 proceedings of PICMET '11*, pp. 1–9.

U.S. Small Business Administration. (2001). *Quarterly indicators: The economy and small business.* Retrieved August 5, 2005, from http://www.sba.gov/index.html

Vesper, K. H. (1980). *New venture strategies.* Englewood Cliffs, NJ: Prentice-Hall.

Watson, T. (2009). Entrepreneurial action, identity work and the use of multiple discursive resources: The case of a rapidly changing family business. *International Small Business Journal, 27*(3), 251–274.

Windsor, D. (2006). Using writing to structure agency: An examination of engineer's practice. *Technical Communication Quarterly, 15*(4), 411–430.

CHAPTER 3

Establishing Program-Specific Assessment Standards for Experiential Learning Courses

Kenneth R. Price

ABSTRACT

While technical communication faculty are in general agreement that learning in professional contexts is and should be a core component of any professional writing program, there are no national standards of assessment for experiential learning courses. This entry outlines a justification and a methodology for using course content guides to establish instructional goals and outcomes assessment standards for experiential learning courses. These guides ensure experiential learning courses best serve the technical communication program, the university constituents, and most importantly, the students. The chapter concludes by explaining how these standards may also be used to anticipate programmatic assessment and verification of experiential learning courses during the institutional accreditation process.

INTRODUCTION

Technical communication programs often rely heavily on experiential learning through internships, cooperative educational experiences, and service learning. In courses with an experiential-learning component or activities, students are able to interact with real-world audiences (i.e., individuals they can expect to meet outside of the classroom) who can provide feedback on their work. While such situations can provide students with valuable input and guidance from outside sources, and can serve as a valuable mechanism for academy-industry interaction, they can be problematic in terms of assessment.

Numerous individuals in the field have advocated for standards for programmatic as well as course assessment in professional communication (Allen, 1993; Carliner, 2003; Hundleby & Allen, 2010; Johnson & Elliot, 2010; Orr, 2010); however, no widespread standards for assessment within the field exist. This situation can, in turn, create problems when it comes to experiential learning situations. Because most of the instruction in experiential learning courses occurs outside of the classroom (and outside of academia) assessment of such courses is particularly difficult. This context is further complicated by the fact that experiential learning courses typically have vague instructional goals, learning outcomes that are difficult to measure, and little or no coordination with industry supervisors (Little, 1993; Sapp & Zhang, 2009; Williams, 2010). Yet such courses remain a valuable mechanism for bringing together academy and industry participants within educational contexts. Effective methods of assessing such experiences can thus play an important role in fostering and maintaining such relationships through experiential learning.

One method of effective assessment of experiential learning courses is through the use of program-specific course-content standards. These standards outline the course's policies, curricular design, methods of evaluation, and far more importantly, detail the course's learning goals, expected outcomes, and methods of assessment. These standards can then be disseminated to all involved with experiential learning activities (e.g., industry partners or internship supervisors as well as technical communication faculty) and thus create a process that allows members of the academy and industry to interact in meaningful ways that can benefit all parties.

REVIEW OF THE LITERATURE

Instructional goals and outcomes assessment has become increasingly important at universities and colleges (Allen, 1993, 2004; Chrysler & Van Auken, 1999; Hundleby & Allen, 2010; Mitri, 2003; Salegna & Bantham, 2002; Taylor, 2002). This focus has, in many ways, become the preferred form of assessment for technical communication courses: "The most sophisticated forms of assessment focus on student learning outcomes: what students learn and what they are capable of doing as a result of their educational program" (Allen, 2004, p. 96). Within this context, the idea of constituents becomes important. As Jo Allen (1993) contends in her review of the literature of assessment, with no national standards, programmatic and course assessment should be based upon how well a program serves its constituents: "The diverse approaches to designing a technical communication curriculum also allow colleges within any given region to provide the kind of program/courses that meet that region's needs" (p. 370). For example, writing for the computer industry could be emphasized in a region with a substantial high-tech industry, or medical writing could be stressed in an area with medical facilities.

While this research suggests college- and program-specific learning outcomes-assessment standards, the difficulty arises of how best to determine, then measure, the technical communication needs of the university constituents. One answer is by having industry supervisors, a key constituent group, understand and be involved in the assessment process of the internship courses. In this way, the assessment process becomes a mechanism for creating and maintaining ongoing relationships or partnerships between members of the academy and individuals from industry/local businesses.

Problems Arising From the Absence of Assessment Standards

With an experiential learning situations having both academic and industry contributors or supervisors, measuring how well such experiences meet their learning goals and expected outcomes can be particularly difficult (Allen, 2004; Allen, Mims, Roberts, Kim, & Ryu, 2006; Furco, 1996; McPherson, 1998; Munger, 2006; Tovey, 2001). Because of this situation, systemic problems appear in the administration, coordination, and assessment of experiential education courses. Such problems can be particularly acute when it comes to situations such as internships. As Sherry Burgus Little (1993) explains, with no set standards, there exists a wide variance in the understanding of the internship itself as well as problems in its planning, supervision, and evaluation. In addition, experiential learning courses are often supervised by individuals who are both outside of academia and thus have no knowledge of programmatic or course assessment. Academia compounds this problem by requesting highly subjective indirect assessments, which attempt to quantify thoughts and opinions about student knowledge and skills, as well as characteristics such as dependability, attitude, adaptability, initiative, and good grooming (Freedman, Adam, & Smart, 1994; Sapp & Zhang, 2009; Williams, 2010). These indirect assessments often involve 5-point Likert-type scale questions that measure opinion—the internship supervisor's perception of what the student has learned. However, verifiable instructional goals and learning outcomes measured by direct assessments, such as those used by technical communication faculty and program administrators, are more effective and useful.

Although these indirect assessments can be valuable in measuring perceptions, values, and attitudes, in the absence of direct evidence, questions arise about how well these assessments measure actual student learning (Newcomer, 2005). Moreover, standards vary from industry supervisor to industry supervisor:

> It is unclear, for example, whether or not a student receiving a low evaluation for the category of writing skills, in fact, has poor writing skills or whether the student merely did not meet the particular industry supervisor's standards for technical communication skills. (Sapp & Zhang, 2009, p. 284)

As a result, this wide variation in standards can lead to problems in maintaining academy-industry relationships.

The problem in assessment is even more pronounced when the internship supervisor is completely removed by the assessment process; an all-too-common problem in which the internship ceases being a partnership (Sapp & Zhang, 2009). For this reason, extra-academic participants in experiential learning projects (e.g., the internship, cooperative learning, and community-service supervisors) need to partner with faculty in assessing the students who participate in such projects:

> As faculty advisors for internship courses, administrators of academic programs, or faculty who teach business writing and communication, we have a responsibility to assess our students' internship experiences to confirm that defined learning objectives are being met, that students' experiences are consistent with our programs' mission. (Sapp & Zhang, 2009, p. 277)

There also needs to be a systematic evaluation of all aspects of experiential learning activities. For example, internship supervisors/industry partners and faculty liaisons should work together to ensure that the related experience is useful and that the quality and quantity of the students' work meets or exceeds the course's instructional goals and learning outcomes.

APPLICATION OF IDEAS

One way to standardize assessment in experiential learning courses and activities is to model industry in having written standards, similar to house style guides. Companies, particularly those that are larger and have multiple writers, editors, proofreaders, and outside developers, typically use a house publications style guide such as the *Microsoft Manual of Style for Technical Publications* or *The IBM Style Guide*. In addition to providing consistency among document sets, style guides and editorial style sheets are accessible to all writers and editors in a company. As a result, these guides or style sheets are often used as definitive or authoritative documents to resolve conflicts by providing the final word for stylistic choices and conventions. Moreover, industry conventions for resolving conflict such as these may provide a foundation for building academy-industry collaborations. In terms of academy-industry relationships, the key becomes using approaches similar to this to guide assessment practices for all involved parties.

Course Content Guides

Course Content Guides (CCGs) for technical communication courses provide a concise description of a particular course, its topical areas, instructional goals, expected student outcomes, and methods of evaluation and assessment. Program- and course-specific assessment standards for experiential learning courses are particularly necessary in the following cases:

- When there are outlying campuses with technical communication courses taught by adjunct faculty,
- When there is little or no coordination of a technical writing program,
- When graduate teaching assistants teach entry-level "service" technical writing courses,
- When programs are small or when departments have few technical writing faculty who have an academic background in technical communication or rhetoric, and
- When a program administrator lacks sufficient credentialing. (Allen, 1993)

Within an experiential learning context, these assessment standards are so important because they are similar to a convention industry professionals generally understand: writing standards. Like a house style guide, CCGs provide all individuals involved in the experiential learning course with agreed-upon standards and provide program, department, and university assessment and institutional accreditation committees with data to assess whether the course meets the instructional goals and educational outcomes constructed from Benjamin Bloom's taxonomy of learning objectives (Bloom, 1956; Bloom, Mesia, & Krathwohl, 1964). Table 1 illustrates the cognitive domain verbs for constructing CCGs.

Because an experiential learning course is typically the capstone course taken at the end of the students' coursework, the instructional goals and learning outcomes of its CCG should be drawn from the three learning domains of Analysis, Synthesis, and Evaluation. These three domains focus primarily on the application of knowledge and skills taken from their technical communication classes:

- **Analysis**: Breaks material or concepts into component parts so that its organizational structure may be understood. Distinguishes between facts and inferences. An example of this category would be writing a troubleshooting section of a training manual or interviewing a subject-matter expert, then selecting the required tasks to create training material.
- **Synthesis**: Builds a structure or pattern from different elements. Revises a process to improve an outcome. Examples of this category would be writing or revising an operational or process manual or designing an object to perform a specific task.

Table 1. Benjamin Bloom's Cognitive Domain Verbs

Course level	Verbs for CCG student outcomes			
100–199	**Knowledge**—Recall information.			
	Arrange	List	Recall	Select
	Define	Memorize	Relate	State
	Duplicate	Name	Repeat	Underline
	Identify	Order	Reproduce	
	Comprehension—Interpret information in one's own words.			
	Classify	Express	Recognize	Simulate
	Cite examples	Identify	Report	Sort
	Describe	Indicate	Restate	Tell
	Determine	Locate	Review	
	Discuss	Practice	Select	
200–299	**Application**—Use knowledge or generalization in a new situation.			
	Choose	Interpret	Restate	Sketch
	Discuss	Pick	Review	
	Dramatize	Report	Schedule	
300–399	**Analysis**—Break down knowledge into its component parts and show relationships among parts.			
	Abstract	Debate	Identify	Outline
	Analyze	Deduce	Illustrate	Predict
	Appraise	Determine	Infer	Question
	Calculate	Develop	Initiate	Relate
	Categorize	Diagnose	Inspect	Review
	Compare	Differentiate	Evaluate	Select
	Conclude	Distinguish	Examine	Separate
	Contrast	Edit	Experiment	Solve
	Correlate	Estimate	Generalize	Test
	Criticize	Examine	Inventory	

Table 1. (Cont'd.)

Course level	Verbs for CCG student outcomes			
400–499	**Synthesis**—Bring together parts of knowledge to form a whole and build relationships for new situations.			
	Advise	Coordinate	Organize	Review
	Arrange	Create	Plan	Rewrite
	Assemble	Design	Predict	Set up
	Categorize	Develop	Prepare	Summarize
	Collect	Devise	Produce	Supervise
	Combine	Establish	Propose	Synthesize
	Communicate	Explain	Rearrange	Systematize
	Compile	Formulate	Recognize	Tell
	Compose	Gather	Recommend	Update
	Construct	Manage	Reconstruct	Write
	Contribute	Modify	Relate	
	Evaluation—Make judgments on basis of given criteria.			
	Appraise	Describe	Measure	Select
	Assess	Discriminate	Monitor	Summary
	Argue	Estimate	Predict	Support
	Choose	Evaluate	Rate	Validate
	Compare	Explain	Relate	Value
	Conclude	Interpret	Review	
	Contrast	Judge	Revise	
	Criticize	Justify	Score	

- **Evaluation**: Includes making judgments about the relative value of information. An example of this category would be selecting the most effective solution to a problem.

These domains can provide a foundation for assessment practices that involve members of both academia and industry by giving both sets of individuals a common foundation for considering experiential learning situations.

Assessment Methods

The use of multiple assessment methods provides converging evidence of student learning. Evaluative methods for experiential learning courses in particular should use direct, indirect, quantitative, and qualitative measurements of the instructional goals and learning outcomes:

- **Direct Measurements**: Standardized pretests and posttests, graded assignments related to learning outcomes, portfolios, and oral examinations.
- **Indirect Measurements**: Student self-evaluations, employer evaluations, directed reflection, or job placement.
- **Quantitative Measurements**: Data that can be measured, such as scores on quizzes and examinations.
- **Qualitative Measurements**: Data that can be observed and described but not measured, such as initiative, maturity, or willingness to take direction.

In relation to experiential learning, these options can provide academy and industry participants with methods that can guide assessment practices and that can include both parties.

Student Self-Evaluations

Student self-evaluations typically involve course exit surveys. The questions for these surveys should be drawn from the course learning objectives. Research suggests that student self-evaluations alone are not a reliable predictor of student ability (Newcomer, 2005). The main problem with relying entirely on self-assessments is validating whether students have actually attained a particular learning outcome. As a result, self-reflections must be supplemented with other means of assessment. Such assessment mechanisms can however provide data useful for gaining a student perspective of the experiential learning class and its administration within the academia-industry partnership.

Standardized Pretests and Posttests

A pretest is a preliminary test to determine a student's baseline knowledge for a course of study. Pretests are particularly effective in helping educators assess whether the content of a course has already been mastered. Often used in conjunction with a pretest, a posttest is given to students after the completion of a course of study to assess the effectiveness of the course. To ensure students are familiar with the seminal concepts of technical communication, questions for pretests and posttests should be standardized and drawn directly from the learning outcomes outlined in the CCG. In terms of creating or maintaining academy-industry relationships or partnerships, the data gathered in these tests can facilitate discussion for continuous improvement of the internship experience.

Supervisor Evaluations

Supervisor evaluations typically take place twice during the semester: at mid-term and the penultimate week of classes. These evaluations can take the form of e-mails, face-to-face meetings, or formal written assessments, such as those illustrated in Table 2. These evaluations provide representatives from industry with a rubric to evaluate student work, while they give faculty quantitative data to aid in assessment. In addition, they provide students with invaluable feedback in the areas deemed important by both industry and academia.

Prerequisite and Corequisite Courses

Prerequisite courses are those required prior to taking an advanced course. Corequisite courses are those that must be taken at the same time as a particular course. Because many students find it difficult to develop the professional skills that they will be evaluated on by industry supervisors (Tovey, 2001), the internship or cooperative learning class may have a prerequisite or corequisite course designed specifically to help students connect their coursework to their internship experience. These courses may include case studies and/or activities to provide the opportunity for applying technical communication knowledge and skills (Sapp & Zhang, 2009). In terms of creating a mechanism for fostering academy-industry interactions, these courses are most effective if designed with the involvement of industry partners.

One of the most effective methods of assessment is directed reflection that contextualizes the internship experience. In fact, many service learning or experiential learning courses require that students reflect upon the experience in writings, exercises, discussions, and readings to place the service work into learning contexts (Moore, 2010). This method of assessment is considered particularly effective because it reframes the learning experience. It is also a method of fostering academy-industry relationships because it further encourages the transition from student to professional.

Although this method is generally highly effective in terms of assessment, it is also has certain limitations. In particular, it is also the method of assessment to which students seem to have the strongest aversion:

> For one thing, students often resist it: They tend to care more about doing the work than about reflecting on it; and they often see the internship as a mode of career exploration, as a foot in the door, and not primarily as a learning experience. (Moore, 2010, p. 10)

Fortunately, a concurrent seminar that offers guided reflection seems to be one way to lessen the students' antipathy to this method of assessment. Accordingly, if both the industry partner and faculty member/advisor closely coordinate the

Table 2. Supervisor Evaluations

Midsemester evaluation	Final evaluation
1. The student understands professional expectations and acts in a professional manner. ____ Disagree ____ Strongly disagree ____ Agree ____ Strongly agree ____ Neither agree nor disagree	1. The student demonstrated the ability to manage multiple priorities efficiently and effectively. ____ Disagree ____ Strongly disagree ____ Agree ____ Strongly agree ____ Neither agree nor disagree
2. The student is reliable and punctual. ____ Disagree ____ Strongly disagree ____ Agree ____ Strongly agree ____ Neither agree nor disagree	2. The student worked independently without constant supervision. ____ Disagree ____ Strongly disagree ____ Agree ____ Strongly agree ____ Neither agree nor disagree
3. The student is able to think and work independently. ____ Disagree ____ Strongly disagree ____ Agree ____ Strongly agree ____ Neither agree nor disagree	3. The student demonstrated effective technological literacy (ability to learn and use computer programs). ____ Disagree ____ Strongly disagree ____ Agree ____ Strongly agree ____ Neither agree nor disagree
4. The student works well with others (collaboration). ____ Disagree ____ Strongly disagree ____ Agree ____ Strongly agree ____ Neither agree nor disagree	4. The student maintained a sense of responsibility for a task or project until completion. ____ Disagree ____ Strongly disagree ____ Agree ____ Strongly agree ____ Neither agree nor disagree
5. The student is willing to accept suggestions, directions, and criticism. ____ Disagree ____ Strongly disagree ____ Agree ____ Strongly agree ____ Neither agree nor disagree	5. The student produced a product useful to the organization. ____ Disagree ____ Strongly disagree ____ Agree ____ Strongly agree ____ Neither agree nor disagree
6. The student demonstrates effective written and oral communication skills. ____ Disagree ____ Strongly disagree ____ Agree ____ Strongly agree ____ Neither agree nor disagree	

experiential learning event(s), then this directed reflection approach can be one of the best methods of gathering assessment data on such learning experiences.

VERIFICATION AND VALIDATION

To become a valid verifiable tool for assessment, the experiential learning course should be structured directly from a detailed curricular guideline like the sample CCG outlined here. Stakeholders should be identified at the onset, and they should be involved in both initially identifying programmatic educational objectives and in the later evaluation of these objectives. The course standards ensure that all the stakeholders involved in experiential learning projects—students, faculty, and industry supervisors/partners—understand the learning goals and outcomes of such projects. These standards also ensure these stakeholder groups can knowledgeably participate in the assessment based upon the benchmarks established at the programmatic level by members of the technical writing faculty or the program director.

What follows is an example rubric that includes some of the more common of the standards that can help forge effective academy-industry partnerships through different kinds of experiential learning courses or activities as well as facilitate effective assessment of such courses or experiences:

I. Course Number and Title—An internship, cooperative education, or other kind of experiential learning course is typically a junior- or senior-level capstone course. For example, ENGL 495: Internship in Professional Writing.

II. Course Description—Advanced application of scientific and technical communication skills in professional/industrial settings. Through their experiential learning or internship experiences and academic coursework, students develop and practice professional skills and master the writing styles for the discipline and the profession. A sample of such a description (for an internship) is as follows:

> The Internship in Professional Writing course enables students to acquire practical experience in a variety of professional settings that draw upon the writing, editing, and design skills that professional writing courses develop. It also enables students to test the practical range of some of the ideas presented in their coursework and to experience firsthand some of the career possibilities available to professional writers and editors.

III. Suggested Text(s)—Written by industry professionals, texts such as these assist in the transition from the classroom to the workplace. Such a listing of suggested texts might include the following titles or works similar to them in scope and focus:

Cram, C. (2009). *New perspectives, portfolio projects for business communication.* Boston, MA: Course Technology.

Hargis, G. (2004). *Developing quality technical information: A handbook for writers and editors.* Upper Saddle River, NJ: Prentice Hall PTR.

Smith, E., & Barnhardt, S. (1997). *Writing at work: Professional writing skills for people on the job.* Lincolnwood, IL: NTC.

IV. Recommended Readings—These seminal texts provide the pedagogical and theoretical foundation of experiential learning situations such as internships, cooperative education, and service learning and/or serve as readings that place such approaches into historical, sociological, and political contexts. These recommended texts could include, but not be limited to, readings such as the following:

Inkster, R. P., & Ross, R. G. (1995). *The internship as partnership: A handbook for campus-based coordinators and advisors.* Raleigh, NC: National Society for Experiential Education.

Kendall, J. C. (1986). *Strengthening experiential education within your institution.* Raleigh, NC: National Society for Experiential Learning.

Linn, P. L., Howard, A., & Miller, E. (Eds.). (2004). *The handbook for research in cooperative education and internships.* Mahwah, NJ: Lawrence Erlbaum Associates.

Little, S. B. (1993). The technical communication internship: An application of experiential learning theory. *Journal of Business and Technical Communication, 7*(4), 423–451.

Qualters, D. M., & Wehlburg, C. (2010). *Experiential education: Making the most of learning outside the classroom.* San Francisco, CA: Jossey-Bass.

Rehling, L. (2000). Doing good while doing well: Service learning internships. *Business Communication Quarterly, 63*(1), 77–89.

Sapp, D. A., & Zhang, Q. (2009). Trends in industry supervisors' feedback on business communication internships. *Business Communication Quarterly, 72*(3), 274–288.

Savage, G. J. (1997). Doing unto others through technical communication internship programs. *Journal of Technical Writing and Communication, 27*(4), 401–415.

Worley, R. B. (2000). Internships in business communication. *Business Communication Quarterly, 63*(1), 62–63.

Although faculty may consult works such as these when setting up an internship class or program, industry representatives familiar with them can play a key role in implementing the ideas into experiential learning courses.

V. Course Design—This section outlines the specifications for the average experiential learning course. Included in this outline are factors such as the number of hours the student is expected to be in a given workplace each week, the academic term(s) the course is offered, and any prerequisite or corequisite courses.

1. Designed to give junior or senior technical communication majors or minors the opportunity to apply technical writing and editing skills in a

professional work setting. Students spend the semester at a job site with a field supervisor providing written assignments and a faculty liaison overseeing both the quantity and quality of written output.

2. Number of course credits: 1–3
3. Laboratory hours per week: 3–9
4. Standard semester time frame
5. Prerequisite(s)/corequisite(s): TCOMM 495: Reflection in Service-Learning Technical Communication Internships
6. Close coordination with industry partners and internship supervisors in English Departments of all outlying campuses

VI. Course Outline or Class Sequencing—In the absence of a true syllabus for an experiential learning project, the outline presented here can serve as a blueprint for internship directors, industry partners/industry supervisors, and students.

1) Complete internship contract
   a. Amount and type of work required
   b. Specific projects to be completed
   c. Firm due dates
   d. Frequency of communication and feedback
2) Complete orientation report and confidentiality agreement
   a. Understand the goals and objectives of the organization
   b. Understand the personnel and organization of the work site
   c. Analyze writing produced by the organization for different audiences
3) Understand the writing needs of the organization
   a. Recognize effective, clear, and technically accurate writing for the organization
   b. Understand and differentiate the conventions and options commonly used by the organization in the presentation of written material
4) Develop a dialogic relationship with the organization
   a. Communicate effectively on the job site
   b. Determine written tasks that are needed and may not be seen yet by the organization
   c. Seek and incorporate critical analysis into writing
5) Apply writing skills to a professional work setting
   a. Tailor writing style to meet the objectives and to follow the conventions of the organization
   b. Produce useful goal- and deadline-oriented written work
   c. Adapt the writing to different audiences
6) Maintain appropriate materials for evaluation
   a. Keep log, journal, and portfolio of all work
   b. Communicate with faculty director and job supervisor on a regular basis

7) Reflection
    a. Submit two reflective papers or journal entries each week to faculty supervisor
    b. Submit at least two written samples of work produced in internship
    c. Compose a longer narrative of the work at the end of the semester
    d. Draft job application materials that include references to the work the students did in their internships
8) Present materials
    a. Oral presentation of the longer narrative of work presented to the faculty director and others who may be invited, with the job supervisor present

VII. Evaluation—Effective experiential learning courses use multiple methods of assessment from multiple individuals. The information listed here provides the various stakeholders involved in experiential learning projects with a mechanism for doing assessment according to various methods used by the different individuals involved (and, ideally, reflect the kind of assessment information each party wishes to obtain). Such approaches include direct, indirect, quantitative, and qualitative measurements of the instructional goals and educational outcomes for experiential learning activities.

a. Course graded A–F
b. Students are graded by midsemester and final evaluations from industry supervisors for semester-long writing and editing projects.
c. The faculty liaison will present the course outline, explain his or her specific policies at the beginning of the semester, and have the student sign an internship contract that includes those policies.
d. Students may be required to submit an orientation report, log book (time and activity), progress reports, directed reflection journal entries, an exit exam, and an oral presentation of a theoretically justified portfolio of projects. May be a community engagement/service-learning internship.
e. Because a key component of experiential learning is reflection, students are required to draft a self-evaluation of work completed.

VIII. Assessment Planning Matrix—The faculty director and industry supervisor should be familiar with this section of the Course Content Guide, particularly the Instructional Goals, Learning Outcomes, and Methods of Evaluation sections. Within this context,

- The **Instructional Goals section** details what the instructor intends to accomplish, as well as what the instructor expects the students to learn in the course.

- The **Student Outcomes section** identifies what the student should know and be able to do as a result of completing the course. These relate to the knowledge, skills, and behaviors that students acquire as they progress through the program. Since the student outcomes must be measurable, the evaluation methods must be able to assess the accomplishment of each outcome.
- The **Methods of Evaluation section** details the exercises, exams, and assignments used to evaluate and collect data regarding student outcomes and programmatic educational objectives (see Table 3).

Additionally, industry supervisors who participate in assessment may provide a wealth of information in creating and maintaining the curriculum guide.

ANTICIPATING INSTITUTIONAL ACCREDITATION

In addition to providing a blueprint for teaching the course, the CCG establishes a baseline for assessment of programmatic effectiveness during the institutional accreditation process (Allen, 2004; Cunningham & Omolayole, 1998; Nichols & Nichols, 2000). A program or institution must have a process in place for internal assessment, and the CCG provides such a process:

- It ensures the assessment process is transparent, publicly documented, and aligned with the program's mission statement.
- It helps to determine how well assessment is linked to curricular practices.
- It helps to identify a program's strengths and weaknesses and allows faculty to work toward continuous improvement.
- It provides outcomes assessment data drawn from both direct and indirect measurements of student learning and program effectiveness.
- It confirms that assessment is ongoing and systematic at the program level.

In sum, academy-industry partnerships established in experiential learning classes may continue in the accreditation process as well. Workplace professionals may share information, evaluation tools, and expertise on issues such as accreditation standards for students' ability to communicate effectively.

CONCLUSION

In this chapter, I have discussed how, in the absence of national assessment standards, experiential learning courses in particular are difficult to assess. In so doing, I proposed course content standards detailing the instructional goals, educational outcomes, and methods of assessment for all technical communication courses and that these be drafted using Benjamin Bloom's taxonomy of learning objectives. In terms of applying the ideas presented here, I would suggest

Table 3. An Example CCG for an Internship Class

Instructional goals	Learning outcomes	Methods of evaluation
Instructional goals indicate the specific characteristics students should exhibit in order to demonstrate desired attainment of the learning outcomes. These goals should be measurable during the evaluation and assessment phases of the class.	Learning outcomes describe what students are expected to know and be able to do at the conclusion of the class. These relate to the knowledge, skills, and behaviors that students should acquire as they progress through the class.	The following exercises and assignments may be used to evaluate what students have achieved or accomplished in the instructional goals.
Develop workplace literacies within the context of a social perspective: collaboration, authorship/ownership, and ethics, as well as the legal issues involved with technical writing, including copyrighting, types of and proper use of company trademarks, and confidential/proprietary documents.	Work independently within the collaborative framework of projects. Be able to adapt writing to different audiences and purposes.	Logs, journals, and supervisor evaluations
Construction print and digital publications that respond to a particular rhetorical situation.	Analyze rhetorical writing situations critically. Understand the concepts and techniques for creating, evaluating, and editing print and digital information.	Coordinated print or digital publication

Revise technical material using standard editing practices.	Evaluate and revise print and electronic technical documents. Seek and be able to incorporate critical analysis into written work.	Editing entry in portfolio
Learn strategies for planning, analyzing purpose, gathering data, drafting, and revising proposals.	Identify and collect data from target populations and stakeholder groups. Describe the various stages of development of documents whose purpose is to fund projects.	Business or grant proposal
Integrate theories and techniques for planning, designing, and writing print and electronic documents using publication standards.	Tailor writing style to follow the conventions of an organization. Identify the various stages of the production process of electronic and print publications, as well as the purpose and components of document- or project-specific style guides and style sheets.	House style guide, editorial style sheets, and word lists
Effective oral communication skills.	Prepare oral presentation, design slides, and handouts. Deliver and connect with an audience: effective oral presentation preparation, design, and delivery.	Presentation of portfolio

that the educational objectives of an experiential learning project should be aligned with a program mission statement and include multiple methods of assessment to measure each educational outcome.

In addition, I suggest all involved in the experiential learning course—faculty member, student, and industry supervisor—should closely adhere to these standards because without this involvement, assessment is weakened, and student learning suffers as a result. I have also shown how these content standards may help to determine programmatic quality and effectiveness during the institutional accreditation process. Most importantly, the approaches and ideas I discuss in this chapter may help to establish or strengthen academe-industry relationships and partnerships.

REFERENCES

Allen, J. (1993). The role(s) of assessment in technical communication: A review of the literature. *Technical Communication Quarterly, 2*(4), 365–388.

Allen, J. (2004). The impact of student learning outcomes assessment on technical and professional communication programs. *Technical Communication Quarterly, 13*(1), 93–108.

Allen, S., Mims, C., Roberts, S., Kim, B., & Ryu, J. (2006). Internship experience: Engaging in the big discourse. *TechTrends, 48,* 44–48.

Bloom, B. S. (1956). *Taxonomy of educational objectives, handbook I: The cognitive domain.* New York, NY: David McKay.

Bloom, B. S., Mesia, B. B., & Krathwohl, D. R. (1964). *Taxonomy of educational objectives.* New York, NY: David McKay.

Carliner, S. (2003). Characteristic-based, task-based, and results-based: Three value systems for assessing professionally produced technical communication products. *Technical Communication Quarterly, 12*(10), 83–100.

Chrysler, E., & Van Auken, S. (1999). A methodology for alumni assessment of an MIS program. *The Journal of Computer Information Systems, 39*(4), 33–39.

Cunningham, B., & Omolayole, O. (1998). An assessment-oriented syllabus model for business courses. *The Journal of Education for Business, 73*(4), 234–240.

Freedman, A., Adam, C., & Smart, G. (1994). Wearing suits to class: Simulating genres and simulations as genre. *Written Communication, 11*(2) 193–226.

Furco, A. (1996). Service-learning: A balanced approach to experiential education. *Expanding Boundaries: Service and Learning* (pp. 2–6). Washington DC: Corporation for National Service.

Hundleby, M. N., & Allen, J. (2010). *Assessment in technical and professional communication.* Amityville, NY: Baywood.

Johnson, C. S. & Elliot, N. (2010). Undergraduate technical writing assessment: A model. *Programmatic Perspectives, 2*(2), 110–151.

Little, S. B. (1993). The technical communication internship: An application of experiential learning theory. *Journal of Business and Technical Communication, 7*(4), 423–451.

McPherson, B. (1998). Student perceptions about business communication in their careers. *Business Communication Quarterly, 61*(2), 68–79.

Microsoft Corporation. (2012). *Microsoft manual of style for technical publications, fourth edition.* Redmond, WA: Microsoft.

Mitri, M. (2003). A knowledge management framework for curricular assessment. *The Journal of Computer Information Systems, 43*(4), 15–24.

Moore, D. T. (2010). Forms and issues in experiential learning. *New Directions for Teaching and Learning, 2010*(124), 3–13. Wiley Periodicals.

Munger, R. (2006). Participating in a technical communication internship. *Technical Communication, 53*(3), 326–338.

Newcomer, J. (2005). Course self-assessments and performance: How well do they correlate? *Proceedings of the 35th ASEE/IEEE Frontiers in Education conference* (p. T3C). Indianapolis, IN.

Nichols, J. O., & Nichols, K. W. (2000). *The departmental guide and record book for student outcomes assessment and institutional effectiveness* (3rd ed.). New York, NY: Agathon.

Orr, T. (2010). Assessment in professional communication. *IEEE Transactions on Professional Communication, 53*(1), 1–3.

Salegna, G. J., & Bantham, J. H. (2002). Curriculum assessment—A systems approach. *Quality Progress, 35*(2), 54–59.

Sapp, D. A., & Zhang, Q. (2009). Trends in industry supervisors' feedback on business communication internships. *Business Communication Quarterly, 72*(3), 274–288.

Taylor, J. H. (2002). Using assessment to build a culture of improvement. *Research Library, 5*(2), 38–39.

Tovey, J. (2001). Building connections between industry and university: Implementing an internship at a regional university. *Technical Communication Quarterly, 10,* 225–239.

Williams, J. (2010). Evaluating what students know: Using the RosE portfolio system for institutional and program outcomes assessment tutorial. *IEEE Transactions on Professional Communication, 53*(1), 46–56.

SECTION II

Industry Contexts

CHAPTER 4

Creating Bridges with Internships

Susan M. Katz

ABSTRACT

It is well known that internships are beneficial for students who need to gain workplace experience to enhance their ability to find a job after graduation, and it is easy to understand the benefits for the host organizations. However, less attention has been paid to the role that internships can play in creating bridges between the academy and local industries. In this chapter, I describe the internship program in the English Department at North Carolina State University, highlighting the benefits for students, hosts, and the academy. I also provide details about the academic course that accompanies the internship experience.

INTRODUCTION

For decades, scholars in a wide variety of fields have recognized the value of the internship experience for undergraduate students. In fact, more than 100 years ago, educators were calling for some type of cooperative education "because both industry and the school are coming to realize that neither of them alone can discharge the educational responsibilities society must discharge to its young workers" (Hanus, 1915, p. viii; see also Schneider, 1909, 1912). A review of the literature on internships, however, reveals that most work published on this topic has focused on the benefits internships have for students or internship hosts. Rarely is the benefit to the educational institution highlighted. Similarly, little attention is given to the role that cooperative education (e.g., internships, service learning, and practica) can play in creating a bridge between the academy and the workplace.

This chapter provides an overview of the internship program in the Department of English at North Carolina State University (NC State). The goal of this examination is to show how such a program can be of benefit to all parties and reveal how these programs can establish or improve academy-industry relationships. In the chapter, I will also share information about the content of the internship course my students take concurrent with an internship and explain some of the tasks necessary for an internship program.

REVIEW OF THE LITERATURE

As Jennifer Bay (2006) notes, the field of technical communication is fortunate to have a wealth of research about the role of internships in preparing our students to enter the workplace. In addition to articles such as Roger Munger's excellent 2006 *Technical Communication* article, which gives advice on how to find an internship (for students), how to coordinate an internship (for faculty), and how to sponsor an intern (for hosts), scholars have looked at a wide variety of issues relevant to the internship experience. These include topics such as

- The way internships enable students to function as entry-level employees (Southard, 1988)
- The value of collaboration between the academy and hosts (Little, 1993)
- The socialization of students to their internship environment (Gaitens, 2000)
- The importance of asking students (especially graduate students) to conduct academic research in addition to the experiential component (Henze, 2006)
- The difficulty that students often have transferring their classroom knowledge to their internship tasks (Sapp & Zhang, 2009)
- The benefit of internships to students, faculty, and the community (Tovey, 2001).

The role of internships in our field is so well recognized that Baywood's Technical Communications Series includes Charles Sides and Ann Mrvica's *Internships: Theory and Practice* (2007). (The Sides and Mrvica book describes an internship program begun at Fitchburg State College in the mid-1970s and provides a model for other colleges and universities.)

We can also learn about the value of internships from research done in other fields. In addition to Andrew Ciafalo's (1992) compilation of older articles representing a wide range of disciplines, there are many relatively recent articles that can add insight to our discussion. For example, in talking about internships for English majors, Bay (2006) tells us that the internship experience, and discussions of those experiences in class, helps students figure out what they like or don't like about a particular job. In so doing, such discussions inspire students to think of careers that they would not otherwise have considered and force them to consider ethical issues in the workplace.

In discussing the value of internships for students in the liberal arts in general, Janet Eyler (2009) provides a set of "guidelines for creating high-quality experiential education programs" (p. 30) that could apply to an internship in any field. These guidelines include the need to consider the following issues:

- Relevance to academic goals
- Assessments that demonstrate achievement
- Responsibility for students
- Workplace supervision
- Academic supervision
- Preparation for the experience
- Reflection on the experience (p. 30)

Similarly, Fred Beard and Linda Morton (1999) provide a detailed set of predictors of successful internships for mass communication majors, and these predictors could easily apply to technical communication students:

- Academic preparedness
- Proactivity/aggressiveness
- Positive attitude
- Quality of supervision
- Organizational practices and policies
- Compensation (pp. 42–43)

Beard and Morton also provide a list of criteria for evaluating the successful internship outcomes, and this list could contribute to a post-internship evaluation completed by each student. The criteria include

- Acquisition of technical skills
- Career benefits
- Career focus
- Acquisition of interpersonal skills
- Practical outcomes (p. 45)

Such outcomes closely parallel those many of us have for our technical communication students and thus could serve as a framework that educators in the field could use to assess the pedagogical efficacy of internships.

Other scholars have investigated issues relevant to internship programs such as concern that the students may be used purely to further the internship host's goals rather than educate them about an industry (Davies, 1997; Dodge & McKeough, 2003), the difficulties of creating good assessment tools for interns (McDonough, Rodriguez, & Prior-Miller, 2009), and the role of reflection in adding to the value of the internship experience (Wasonga & Murphy, 2006).

So with all this good information available, what can this chapter contribute to your understanding of technical communication internships and how they foster good relationships between the academy and the university? My hope is that by describing NC State's internship program and the benefits it offers students, hosts, and the university, readers will find a reason to develop an internship program or a way to improve the current internship programs provided to students. Through such reconsideration, readers might also gain new perspectives on the benefits current and prospective internship programs could offer departments or institutions.

METHODS

This chapter provides a case study of my experience working with student interns in the Department of English at NC State. In 1997, Dr. Judi Gaitens asked me to collaborate with her to propose an internship course for English majors at NC State. Such a course would be targeted primarily at students in what was then known as the Rhetoric, Writing, and Language (now Language, Writing, and Rhetoric) concentration within the BA in English, which consisted of students focusing on technical writing, journalism, or linguistics.

Prior to that time, students were allowed to earn academic credit for internships, but there was no academic component to those internships. Dr. Gaitens felt it was important to create a more responsible mechanism for matching students with internship hosts while also monitoring their progress and supporting their transition to the workplace. The proposal was approved, and Dr. Gaitens offered *ENG350: Internships in Writing and Editing* for the first time in the spring semester of 1998. When she retired in fall 2000, I took over teaching the course, and I assumed full responsibility as Internship Coordinator for the program in fall 2005.

Since 2005, I have used pre- and post-internship surveys to collect data from students who have completed internships (although not from every student and not every semester) as well as collected final course evaluations (data collected by the university each semester) for these programs. Additionally, internship hosts submit intern evaluation forms (although, unfortunately, not every host complies with the request for evaluations). Finally, I have anecdotal data in the form of internship-related e-mail messages or handwritten notes from both students and hosts. Excerpts from these various materials will supply evidence used to support my discussion of and conclusions about the role of an internship program in helping to connect the academy and the community.

THE NC STATE DEPARTMENT OF ENGLISH INTERNSHIP PROGRAM

In its current form, the program (which has a required academic component) was created specifically for students who were interested in careers in technical communication or journalism. NC State does not have specific majors in either of those fields; those students earn a generic English degree. The course has been offered every fall and spring semester since 1998, and for the first few years, it served approximately 10 students per semester. The program has since grown to encompass not only all English majors (that is, students with interests in literature, film, linguistics, education, and creative writing in addition to technical communication and journalism) but is also open to all students majoring or minoring in any program in the NC State College of Humanities and Social Sciences (CHASS). For the past five years, the program has averaged 20 students per semester, the majority of whom are either English or communication majors, many of whom minor in journalism.

Initially, the organizations that hosted our interns were primarily newspapers and magazines or private companies interested in technical writers. However, as the interests of our students have expanded over time, so has the list of prospective internship hosts. When I first took over the administration of the program (fall 2005), my predecessor gave me a list of about 30 active internship hosts. In the spring of 2009, I made a concerted effort to expand that list by contacting local businesses and asking to meet with them to talk about the type of work that the students in my program could contribute to their organization. These meetings were significant in that I was able to establish relationships that have been ongoing and valuable, both in terms of creating new internship opportunities and also in that I was able to demonstrate that NC State has excellent students in areas beyond the well-known fields of agriculture and engineering.

I did additional marketing of the program by revising the program website, creating an internship alumni newsletter, and using the resources of the NC State Career Development Center to reach additional employers. As a result of these activities, my program currently has relationships with more than 200 potential internship hosts, primarily in our local area.

In this section, I will describe the requirements, procedures, and course content for NC State's ENG350 Professional Internships[1] before moving on to discuss the benefits the program offers students, hosts, and the university.

[1] The name of the course was changed in fall 2011 when the course was officially opened to all CHASS majors.

ENG350 Professional Internships

Students interested in participating in an internship through the English Department's program at NC State must complete an application process consisting of

- An application form
- A cover letter
- A résumé
- Two writing samples
- Two faculty recommendations

Additionally, students are required to be in their junior or senior year at NC State and have a minimum overall GPA of 2.5, with a minimum GPA of 3.0 in their major.

As Internship Coordinator, I review each application and then contact students to let them know if their application has been approved. I also provide students with suggestions for potential internship hosts aligned most closely with each student's interests. The students are directed to first look for positions on ePACK, the NC State Career Development Center's job and internship database, but they are also encouraged to communicate with me about other possibilities. We try to ensure that every student is placed in an internship prior to the beginning of each semester.

Students who wish to participate in the internship program are required to enroll in *ENG350 Professional Internships*, which meets for 75 minutes once a week and is taught by the Internship Coordinator. In the weekly classroom sessions, participants discuss techniques for success in the workplace, career options for CHASS majors, and job search strategies and skills. These discussions include topics such as résumés, cover letters, portfolios, and interviewing techniques. The students also read articles on relevant topics as well as two books: *35 Tips for Students to Succeed in Corporate America* (Hill, 2007) and *Strengths Finder 2.0* (Rath, 2007).

Throughout the semester, guest speakers visit the class and share their expertise with the interns: Career counselors talk about resources for finding jobs, professional recruiters and human resources managers discuss what they look for in job applicants, and entrepreneurs explain how they got their start. Other guests have careers relevant to CHASS students, and they describe the paths they took to get where they are today and talk about the advantages and disadvantages of their current positions. In selecting guest speakers, I try to include professionals from large corporations and small businesses, nonprofit organizations, and government agencies so that students are exposed to a wide variety of career options.

There are a number of assignments that all registered students must complete. Table 1 provides a concise description of each of these assignments.

Table 1. Student Assignments

Assignment	Description	Value
Cover Letter and Résumé	Find an appropriate advertised position. Respond with a cover letter and résumé. Also submit the ad so that the instructor can evaluate the relevance of the response.	Provides practice in adapting the cover letter and résumé to the requirements of a specific position.
Mock Interview	Complete a mock job interview with one of the Career Development Center's career counselors.	Provides practice in job interview skills.
Strengths Finder 2.0	Read the book. Complete an online assessment that results in identification of five strengths. Review the results of the assessment. Choose two "ideas for action" (from a list provided in the book) for each strength and submit a written description of how the student can incorporate that idea to build on the strength. Also, choose two traits that were not identified as strengths but that would be useful to develop, and submit a written description for how the student could develop those traits.	Asks students to identify and think about their strengths. Helps prepare them for interview questions about strengths and weaknesses. Assessment response also provides career suggestions based on the individual's strengths.
Daily Internship Log	Record hours worked, tasks completed, skills developed.	Provides students with a record of their hours so that they will be able to demonstrate that they have met the course requirement. Also, gives student concrete evidence to use in job applications.
Internship Summary	A two-page reflection on the value of the internship experience.	Provides students with an opportunity to identify what they have gained from the internship experience.
Portfolio	Create a hard copy or electronic portfolio of written work. Can include materials from the internship as well as coursework, other internships or jobs, volunteer experiences, and extracurricular activities.	Gives students a tangible demonstration of their writing skills. Excellent resource for job applications and interviews.

As Internship Coordinator, I also plan a major internship-related event each semester. The purpose of these events is to help the students learn about potential careers. The fall event involves CHASS alumni, and the spring event involves local employers. The students in the internship course are required to attend the event, but it is open to all students and is advertised via electronic billboards, flyers, and newsletters.

The fall event is an alumni/student networking Open House that provides current students with opportunities to chat with alumni who have been successful in launching their careers. Generally, 15 to 20 alumni attend this event, and these alumni include journalists, technical writers, public relations experts, entrepreneurs, managers, trainers, museum workers, and representatives from the nonprofit and government sectors.

The spring event is a Career Networking Event at which local employers come to talk with students about the types of jobs that exist in different areas. Unlike at a job fair, the employers who meet with students do not necessarily have job openings; they can, however, help expand the students' approaches to the job search and help students expand their professional networks.

Once students have completed the Professional Internship course, they are eligible for two additional benefits. First, I will add them to an e-mail list I maintain (as Internship Coordinator) for students and alumni who are interested in learning about job openings that come to my attention. (Because of the bridge-building activities described earlier in this section, many local employers send me information about job or internship positions.) The second additional benefit is that students who want to gain more experience can do a second internship, with a different organization, as an independent study with me. This option allows students to earn academic credit for a second internship. This is an important benefit for students as many hosts insist interns earn academic credit while working within their organizations.

For the independent study, students read assigned articles, submit written reflections on the relevance of those articles to their internship experience, revise their résumés, meet with me (as Internship Coordinator) on a biweekly basis, and complete some type of major project. For most students, the major project is a portfolio of their writing, which they will create electronically if they created a hard-copy portfolio in ENG350 or in hard copy if their ENG350 portfolio was created electronically. A few students have preferred to produce a different type of product, such as a research report on career opportunities within a given field or detailed descriptions of relevant job titles, such as the various jobs that could come under the heading of "editor." For a more specific example, in the summer of 2012, an independent-studies student working in the production department of a local television station chose to write a report looking at the relationship between news and sales departments in the broadcasting industry.

BENEFITS OF AN INTERNSHIP PROGRAM

As described previously, the benefits of internship programs have been discussed by many scholars over the years (see pp. 77–80 of this chapter). In this section, I will provide specific examples of how the internship program in the English Department at NC State has benefited students, hosts, and the academy.

Benefit to Students

Dodge and McKeough (2003) tell us that "internships . . . contribute to the students' ability to put that all important work experience on their resume" (p. 46). While that is certainly true, students recognize that the value of the internship course goes far beyond improving their ability to gain employment after graduation. For the sake of brevity, I will include only one or two examples (from the sources listed in the Methods section) to support each claim.

One of the most common student responses pertains to the usefulness of the Professional Internships course in preparing them for the job search process. In terms of the benefits of the course, one student noted,

> This course is probably the most useful course I have ever taken. From writing resume and cover letters to mock interviews, the course prepared me for entering the job market. I love that we do a writing portfolio because I have found that is important when trying to secure a job since I do not have much experience. (final course evaluation, spring 2009)

And another student noted, "I especially liked hearing from all the guest-speakers, working on a resume and cover letter, and learning so many tips to prepare us for life after college. The mock interview was a great idea too" (final course evaluation, fall 2011).

In other cases, students highlight the benefits provided by the internship experience itself or the internship in combination with the class, as is noted in comments from the postinternship surveys: "My internship prepared me to become a better writer and taught me how to work in a corporate environment efficiently. Working with so many employees taught me how to work alongside of many personalities" (postinternship survey, fall 2011; student worked at a local television station). Another student commented,

> I am most certainly better prepared for a job because I now have valuable experience with the software used to create print and e-books. Before my internship, I was simply prepared to write essays and literary analyses, but now I have practical, marketable skills that could be utilized in any number of professions. (postinternship survey, spring 2010; student worked for an e-book publisher)

In yet other instances, students note the importance or the contributions of a particular assignment or aspect of the class that they found to be important after graduation:

> You'll be pleased to know it was my portfolio for your internship class that really got me the job. They were really impressed with how organized it was. Honestly, that class was one of the best choices I've ever made. (K. Herron, personal communication, November 18, 2010)

Another student noted, "Thanks so much for sending out all these wonderful job opportunities. I am constantly passing them along to friends and family members [who] are still unemployed" (J. Dunkley, personal communication, July 21, 2009).

These comments reveal that well-structured and effectively supervised internship programs that merge academic and workplace experiences have a great deal to offer today's technical communication students. What is difficult to measure, and what may be the most significant benefit to students, is the way that an internship demonstrates for students how to *use* the skills that they have developed in their coursework in a new environment. The course that I teach does not ask students to write journalistic articles or press releases or plan events or complete any of the typical tasks that my interns do at their various workplaces. However, their ability to complete those tasks is dependent upon their prior successful completion of academic assignments—their ability to think about the audience and purpose of specific writing tasks, to meet deadlines, to collaborate with others, and to make presentations (informal and formal) about their work and their processes. Thus, the internship experience builds on the work they have done in their individual curricula and helps them see how to apply that classroom knowledge to workplace situations.

Benefit to Hosts

One of the most obvious benefits to hosts, especially in an era of workforce reductions and economic cutbacks, is the work that the interns do within their organization. Not only is the work highly valued but most hosts feel that the effort they put forth to train the interns is well rewarded. The following excerpt reveals how internship providers value such programs and the students who participate in them:

> Jeannene did an excellent job for us this semester. She was a regular contributor to our blog, including a weekly calendar of events that she took on as a regular assignment with little weekly guidance from the editors at the magazine. She embraced an idea we had for a regular blog story on NC State fans who live near the town . . . of the team NC State was playing in football each week. She provided us photos and other coverage from campus events, extending the reach of our staff. (final performance evaluation, NC State alumni magazine, fall 2011)

Many of the interns in our program are skilled in a variety of areas that fall under the general rubric of technical communication (e.g., technical editing, technical writing, document design, social media writing, etc.). As a result, interns often provide organizations with expertise that their employees may lack. This factor, in turn, provides an added benefit to both the intern (who is able to contribute meaningfully to the sponsoring organization) and the internship provider (who is able to benefit from the addition of this new skillset) as exemplified by the following comment:

> Amber was invaluable in helping us tell the story of Arts, Heritage, and Culture in North Carolina in a number of ways, including writing, research, and giving a young person's eye to social media as well as video production for the North Carolina Awards. She also pitched in as needed on deadline-oriented projects. She was a real asset, and we will miss her next semester! (final performance evaluation, North Carolina Department of Cultural Resources, fall 2011)

Occasionally, an intern will bring an unanticipated approach to the workplace, and such novel perspectives can be a great asset to open-minded hosts who recognize and embrace them. In the fall of 2011, for example, a student who was a creative-writing major gained an internship for a nonprofit organization that conducts research and provides educators with resources pertaining to the North Carolina coastal area. One of the major efforts of this organization is the publication of a monthly magazine, *Coastwatch*, which typically consists of research reports relevant to the coast as well as articles fostering recreation and coastal industries. On the intern's final evaluation, her internship supervisor wrote,

> Tina has worked on varied projects during her semester, with the focus on items for *Coastwatch*. In the holiday issue, she had a number of Tidings and a sidebar to a lead story about Hurricane Irene damage. But I think that her greatest contribution to that issue was her poem[2] on the back cover. It was a departure for the magazine to include staff poetry, but we were inspired by Tina's skill and interest. (final performance evaluation, *Coastwatch*, fall 2011)

Our internship program also assists employers by providing easy access to our students for both internship and full- or part-time positions. As noted above, employers frequently contact me directly when they have employment or internship needs. In these cases, prospective employers demonstrate that they

[2] You can read Tina's poem in the electronic version of the magazine at http://www.ncseagrant.org/home/coastwatch by clicking on the link for the Holiday 2011 issue and then on the title, "In Irene's Wake."

trust us to make recommendations based on their experience with students. For example, the following message came from an employer I had worked with in the past:

> We are recruiting for Public Relations Interns for next semester. We have had very few applicants, and I wondered if you would be willing to share our position description with students and faculty in your department. We have some exciting new projects starting next semester that promise great work experience for anyone interested in public relations and marketing. (P. Lipford, personal communication, December 13, 2011)

In addition to "return clients," I also hear from employers who are new to the area. This message came from an employer who had read about our internship program on our program website:

> We are The Total Connection newspaper and are trying to start a new website called TTC Kids News. We want to create a website with nothing but videos of the kids in the community and what they do based on our newspaper. We are looking for part-time interns to direct, produce, write scripts, film, etc. How do I post this job to your website? (R. Harmon, personal communication, December 7, 2011)

It is also fairly common for alumni to write to me when their organizations have employment needs. The alumni know from personal experience that I, as Internship Coordinator, maintain the "job opportunities" mailing list mentioned above and that this list has and can provide employers with excellent candidates. The following message came from such an alumna:

> Good morning Dr. Katz! I hope this email finds you well. I took your internship class Fall 2007 (and will still say it was one of, if not the most beneficial class I took at NC State.) I have worked at the North Carolina State Employees' Credit Union as an e-learning developer (writing and developing online training for employees of the organization) since graduation. Currently, there is a position available for an e-learning developer at NCSECU in Raleigh. I have attached the job description. I thought you might have some former students on your list-serve who may be interested in the job. (L. Barfield, personal communication, August 15, 2011)

All these comments indicate that internship programs can build a bridge that creates an effective avenue for academy-industry communication. Moreover, comments like these note that internship programs also serve as a bridge students can use to transition successfully from student to employee.

Benefits to the University

Although it is fairly easy to talk about the benefits internship programs offer to students and to hosts, the benefits they offer to a department or a university are significantly less tangible and more difficult to enumerate. As I write this chapter, however, many writing programs in colleges and universities are facing serious questions from administrators and funding organizations (such as state legislatures for public universities) about the value of the education that they offer. Internship programs can be a key response to those threats, demonstrating that the type of skills students learn in the humanities can help them succeed in a variety of careers and find gainful employment upon graduation.

That is (and this is one of the most obvious benefits), a good internship program will (ideally) improve job placement for students after graduation. My department does not keep placement statistics, and it would be difficult to compare the placement of students who have done internships with those who haven't, but recent research tells us that employers are looking for students who have completed internships (Millennial Branding, 2012). Thus, we can suggest that students who have completed an internship will have an easier time finding work than those who have not.

The number of requests I receive from local employers for interns and new hires (from mid-May 2011 to mid-May 2012, I received approximately 150 requests) confirms that my English Department's internship program helps to build the reputation of the department and the university with a variety of private and public organizations across the state and the region. There is also a financial benefit to the university as local internship hosts make good candidates for donations to scholarship funds or other department programs. Although I have not had time to pursue such contributions on my own, I have shared the list of active hosts with the CHASS Office of Development.

Moreover, former interns who do well and who credit their internships with helping them land a good job are often among the biggest individual contributors to the English Department Interns' Scholarship Fund. To date, alumni have contributed nearly one third of all the money collected for this small fund. Some students even contribute before they graduate, as the following message demonstrates:

> This semester I am already realizing how much your [internship] class has helped me—I feel much more confident in interviews, it's great knowing that I have an internship under my belt, thank you, thank you. I would also like to donate a (albeit, small) sum towards the English Internship Scholarship fund. If it wasn't for this scholarship I wouldn't have been able to take your class. And honestly, much has come as a result of this class. Who or what can I make the check out to? (M. Novak, personal communication, March 9, 2010)

Furthermore, hosts assist the department not only by providing internship and job opportunities for our students and graduates, but also by volunteering their time. Many of the guest speakers who come to talk to the students in the internship class or who participate in the annual networking event have previously hosted interns. Through those experiences, these internship providers have come to appreciate the work that we do in the classroom. They also recognize that we are trying to prepare students for their roles within the workplace, and they often offer to speak to students in other classes, to mentor students individually, and to help students network as they begin their job searches.

At NC State, the internship program has also helped to enhance the visibility of our programs and our department in an institution where it is easy to be overlooked based on the sheer size of the organization. One CHASS department head, on walking past a bulletin board displaying flyers for the guest speakers for the internship course, told my department head that he wished he had a program like that in his department. The CHASS Dean has touted the program in conversations with local businesses as a way of encouraging them to consider the value of CHASS students for their organizations and has also used it as a way to engage the active involvement of alumni (either by encouraging them to serve as hosts for interns or contributing to scholarship funds).

As mentioned previously, the internship program helps us maintain a relationship with alumni. Many (if not most) of the alumni who take the time to complete a profile on our department's Alumni and Friends website participated in the internship program as undergraduates or graduate students. Maintaining relationships with alumni is a difficult challenge, so any activity that makes this easier is of value to the university.

Another important benefit is that an internship program helps students who participate recognize the practical application of the courses that they have taken, think more clearly about career opportunities, and prepare for life after college. As mentioned previously, internship experience can help alleviate some of the problems with the transferability of classroom knowledge to the workplace setting simply by giving students an opportunity to practice the use of that knowledge.

We know that students who have completed internships are more marketable, and as those students become solid employees, they help enhance the reputation of the university in the workplace. Many NC State students stay within the local area—a region that is one of the fastest growing areas in the United States—so they contribute to the status of the university through their work.

CHALLENGES

Although an internship program can be a great benefit to all concerned, there are many challenges in creating and maintaining this type of program. In addition, some of the tasks inherent in producing a successful program are unfamiliar ones for college instructors. In this section, I will describe the various activities involved in producing an internship program similar to the one in the English Department at NC State.

Publicizing the Program

Oddly enough, students don't flock to the internship program. It is not required of any English major and does not meet a specific requirement for most of them. Each semester, I create printable flyers, bookmarks, electronic displays (we have electronic bulletin boards throughout the university), and fact sheets for the program. I ask my colleagues to allow me or one of the current interns to speak to the students in their classes about the internship program. In undertaking such activities, I teach the interns to make a 5-minute presentation about the program, and I coordinate the interns' schedules with my colleagues' class schedules, making presentations myself when no intern is available. This approach has traditionally been our most effective marketing strategy.

Establishing (and Maintaining) Relationships with Hosts

As stated earlier, I have met personally with many local hosts, but most of my regular communication is e-mail correspondence and the occasional phone call. This communication can involve answering questions about the program requirements, inquiring about internship opportunities in general or for a specific student, sending agreement or evaluation forms, or sharing information about part- or full-time positions.

Corresponding with Prospective Interns

Over the history of the program, it is typical that about twice as many students will write to inquire about the program as actually end up registering for the course. Students will write to me to ask how an internship might fit into their schedule, what types of internships are available, how to apply, what types of writing samples to provide, what the assignments are within the course, and so forth.

Reviewing Applications

The application and faculty recommendation forms are online and come to me automatically when the writer clicks on the "submit" button, but the required résumé, cover letter, and writing samples are sent via e-mail and often sent piecemeal. Because of this, I have to keep track of what has been submitted to ensure that I eventually have all the required materials. I typically review the materials as they are submitted, as I often need to give students feedback on how to write a résumé or cover letter that would be acceptable to an internship host. It is also not uncommon for me to send a "gentle reminder" to faculty who have not submitted the recommendation form for several weeks after it was requested. I do not add students to the course until the application is complete and the materials are acceptable.

Corresponding with Administrative Assistants

The course is restricted as students must be admitted to the program before they can be enrolled, so as each student is accepted, I notify the student and the administrative assistant. In some instances, this requires several rounds of communication as there could be holds on student accounts, the student may have registered for too many credit hours, or there could be a scheduling conflict.

Matching Students with Internships

Although I ask students to do most of the legwork involved in finding an internship on their own, I still end up providing suggestions and making contacts for them in those cases in which we have never had an internship with a particular host organization in the past.

Preparing to Teach the Course

Having taught the course many times now, the course prep is significantly less once the semester has started. The sessions on topics such as résumé and cover letter writing need only minimal revision, if any, each semester. However, given that I invite anywhere from 12 to 15 guest speakers to talk with students, the preparation prior to the beginning of the semester is fairly intense as I attempt to logically schedule them at appropriate points in the semester.

Although I am fortunate that all of the department heads I have worked with (and my current dean) have valued the program and appreciated my efforts, this is not the type of work that typically gets rewarded by the academy. It counts as either administrative or service work and should not be undertaken by anyone who doesn't already have tenure.

APPLICATIONS TO OTHER CONTEXTS

While this chapter has focused specifically on one internship program within one department of a large public university, the basic concepts that provide the foundation for such a program can be modified and adapted to any context. I am currently working on a large-scale survey of internship programs in other universities, but as a pilot project, I conducted a survey of internship programs in departments and institutions similar to NC State. While I have not yet completed a detailed analysis of the results, I have identified key questions that need to be addressed when creating an internship program and can suggest some variations in how these questions are answered at different institutions.

- *Where will the program be housed?* The program may be housed within a specific concentration (e.g., technical communication), department, college, or in a centralized career center that manages internships for all programs throughout the university.
- *Who will manage the program?* In most cases, someone on the faculty of the sponsoring department manages the program; however, in some instances, the program is managed by a professional academic advisor or career counselor.
- *What are the criteria for participation?* While criteria vary, and some schools ask only that the student find an internship, the most common criteria include prerequisite courses, minimal GPA levels, a specific level of degree completion (e.g., third or fourth year), and/or faculty recommendations. Decisions also need to be made about whether a program will focus exclusively on undergraduates or also be available to graduate students.
- *How does the student find an internship?* Students may be required to find internships for themselves. In most instances, programs at the very least have lists of hosts that have been developed over time, and typically the internship coordinator is available to make suggestions.
- *What is the academic component?* In addition to the option of an actual class, other possibilities include preparation of a reflective report at the end of the internship, regularly scheduled individual meetings with a faculty adviser or the internship coordinator, or an independent study course. These options can be the same or different for undergraduates and graduate students who participate in an internship program.

Another variation involves how much academic credit students can earn. When a course is involved, students typically earn the same number of credits as for any regularly scheduled course within the department. However, when the internship is handled as an independent study, the number of credits can vary depending on the number of hours worked. Some institutions allow students to do multiple internships and earn credit for each of them up to an established maximum.

- *How much time must the student spend at the internship?* Again, this varies from one program to another, but most commonly the students work approximately 40 hours for every academic credit hour that they will receive.
- *How is the internship experience evaluated?* Most programs ask employers to evaluate the work of the student interns and provide that evaluation to the program coordinator. About half of the programs surveyed also ask students to provide an evaluation of their experience.

CONCLUSION

Although an internship program such as the one described requires a significant commitment from the sponsoring department, the benefits outweigh the costs. As outlined above, there are benefits for everyone involved, but in my conclusion I will focus on the benefits of establishing and improving relationships between the academy and local industries.

One of the most important things that an internship program can do for the academy is give faculty a reason to contact their counterparts in the workplace. For example, I have in the past contacted organizations on behalf of students who wanted an internship within a specific industry. Our students in technical communication typically have no difficulty finding internships because we are fortunate to have a number of industry giants locally, including IBM, SAS, and Red Hat. However, many students would prefer to work (and learn) in smaller organizations. As a result, I have reached out to many such companies with technical communication needs and now have a list of these organizations that I can provide to students who want to gain this type of experience.

Another reason to contact employers is to learn more about an organization brought to my attention by a student intern. In the spring 2012 semester, one of my interns (who is not in a technical communication concentration) had found a position on his own working with a training, design, and IT contractor. After corresponding about the intern throughout the semester, the employer asked to meet with me in person for a more detailed discussion of the work the organization does and their potential for hosting additional interns. The face-to-face meeting helped us both recognize that graduate students in NC State's MS in Technical Communication program as well as the students in our PhD program in Communication, Rhetoric, and Digital Media would be excellent candidates for paid internships as well as future part- or full-time employment. Subsequently, they hosted (and paid) a graduate student intern for the spring 2013 semester and hired her part-time for the summer following the internship experience.

In some instances, there are organizations, such as IBM and SAS, with which we have a well-established relationship, but the organization is so large that we are only known in certain departments. I have used internship program contacts (guest speakers who visit my class, internship supervisors, alumni employees) to gain introductions to individuals in other departments within these organizations, thus increasing awareness of both the program and myself as a resource they can tap when they are in the market for either interns or employees. These visits often result in offers from additional employees to speak to my students or to students in other technical communication courses.

The connections that I make in various industries, whether by inviting guests to come speak to my class, outreach on behalf of prospective interns, or my own initiative to learn more about an organization, have multiple benefits. The obvious benefits are to students and employers, but there is also benefit for the courses and programs within my department. My conversations with industry representatives help me learn about their needs, which I can share with my colleagues and can in turn affect what we teach in our classes to better prepare our students for their future careers. However, those conversations also help local employers understand what it is that we in the academy do that may benefit them more than they realize, such as how our research informs our teaching, which then affects the students we produce, making them (ideally) critical thinkers who will be professional, innovative employees.

In closing, I should add that I have found the experience of coordinating the internship program to be the most rewarding aspect of my academic career. The students who participate in the program are genuinely—and effusively—grateful. They send me e-mail and handwritten notes. They stay in touch with me on Facebook and LinkedIn. They report on their successes (and failures). They let me know when their employers have open positions. They continue to ask for advice, even years after graduation. They are the reason I became a college professor.

REFERENCES

Bay, J. (2006). Preparing undergraduates for careers: An argument for the internship practicum. Symposium: What should college English be? *College English, 69,* 134–141.

Beard, F., & Morton, L. (1999). Effects of internship predictors on successful field experience. *Journalism & Mass Communication Educator, 53,* 42–53.

Ciofalo, A. (Ed.). (1992). *Internships: Perspectives on experiential learning.* Malabar, FL: Krieger.

Davies, S. (1997). Internships as a publishing tool. *Journal of Scholarly Publishing, 28,* 227–233.

Dodge, R. B., & McKeough, M. (2003). Internship and the Nova Scotia government experience. *Education + Training, 45,* 45–55.

Eyler, J. (2009). The power of experiential education. *Liberal Education, 95,* 24–31.

Gaitens, J. (2000). Lessons from the field: Socialization issues in writing and editing internships. *Business Communications Quarterly 60,* 64–76.

Hanus, P. H. (1915). Introduction. In H. Schneider (Ed.), *Education for industrial workers: A constructive study applied to New York City.* Yonkers-on-Hudson, NY: World Book.

Henze, B. R. (2006). The research-experiential internship in professional communication. *Technical Communication, 50,* 339–347.

Hill, S. A. (2007). *35 tips for students to succeed in corporate America.* Raleigh, NC: Lulu.com.

Little, S. B. (1993). The technical communication internship: An application of experiential learning theory. *Journal of Business and Technical Communication, 7,* 423–451.

McDonough, K., Rodriguz, L., & Prior-Miller, M. R. (2009). A comparison of student interns and supervisors regarding internship performance ratings. *Journalism & Mass Communication Educator, 64,* 140–155.

Millennial Branding. (2012, May 14). *Millennial branding student employment gap study*. Retrieved May 30, 2012, from http://millennialbranding.com/2012/05/millennial-branding-student-employment-gap-study/

Munger R. (2006). Participating in a technical communication internship. *Technical Communication, 53,* 326–338.

Rath, T. (2007). *Strengths finder 2.0.* New York, NY: Gallup.

Sapp, D. A., & Zhang, Q. (2009). Trends in industry supervisors' feedback on business communication internships. *Business Communication Quarterly, 72,* 274–288.

Schneider, H. (1909). Fundamental principles of industrial education. *Transactions of the American Institute of Electrical Engineers, 28*(1), 269–278.

Schneider, H. (1912). Education and industrial peace. *American Academy of Political and Social Science, Annals, 44,* 119–129.

Sides, C. H., & Mrvica, A. (2007). *Internships: Theory and practice.* Amityville, NY: Baywood.

Southard, S. (1988). Experiential learning prepares students to assume professional roles. *IEEE Transactions on Professional Communication, 31,* 157–159.

Tovey J. (2001). Building connections between industry and university: Implementing an internship program at a regional university. *Technical Communication Quarterly, 10,* 225–239.

Wasonga, T. A., & Murphy, J. F. (2006). Learning from tacit knowledge: The impact of the internship. *International Journal of Educational Management, 20,* 153–163.

CHAPTER 5

Collaborating with Industry Using Mentoring Programs and Internships

Herb J. Smith

ABSTRACT

Mentoring programs and internships are two of the more effective ways to build lasting partnerships with industry. Mentoring programs and internships also extend learning beyond the classroom and allow students to apply what they have learned in the classroom to workplace situations. This chapter provides a description of formal and informal mentorship programs and guidelines on how to set up each. The chapter also examines different internship models and discusses the benefits of including internships in academic programs in technical, scientific, and professional communication. In so doing, the chapter presents an approach others can follow for creating effective academy-industry partnerships and relationships.

INTRODUCTION

For many years, collaborative ventures between the academy and industry were difficult to set up due in part to inherent differences placed on the value of theory versus practice. As Bosley (1995) notes, industry traditionally downplayed the importance of theory over practice, viewing theory as unnecessary, while academicians emphasized theory over practice in the complete education of the individual. Hayhoe (1998) presents a similar viewpoint and argues for a new model of education that promotes both learning and learning to do.

However, in the past 15 years or so, much has changed. For example, mentorship programs are gaining in popularity as a successful way to partner with industry, and internships are widely used as a successful type of experiential learning in many technical, scientific, and professional communication programs bridging theory and academic learning with practice. Successful mentoring and internship programs can help build strong ties between the academy and industry rooted in the Aristotelian concepts of *episteme* (knowledge) and *techne* (craft or art).

This chapter examines this relatively new model of education that Hayhoe (1998) argues for, and the chapter does so by exploring the important role that mentoring programs and internships play in building strong partnerships between the academy and industry. Specifically, this chapter contains the following sections:

- Method of Examination
- Review of Literature
- Origins and Nature of Mentoring
- Different Mentoring Models With Survey Results
- Some Guidelines for Building Effective Mentoring Program
- Student Internships as Gateway Experiences
- Different Internship Program Models
- Internship Benefits
- Application of Ideas
- Conclusion

An examination of these topics offers both those in academia and in industry a more complete understanding of how to engage in effective mentoring practices and how to create internship programs that can foster successful academy-industry relationships and partnerships.

METHOD OF EXAMINATION

Information for this chapter was gathered through several different methods. Information from participant observations was used, particularly student interviews with mentors and sample student internship final projects. An online survey of mentors was also conducted to get their feedback on the mentoring experience. The size of the survey was small (13), with 9 mentors responding. Nonetheless, their responses provide valuable insight into the mentoring experience from the viewpoint of the mentor.

The yearly bibliographies (1999 to 2011) of the Association of Teachers of Technical Writing (ATTW) were reviewed for books and articles pertaining to mentoring and internships, as was the EServer TC library for articles on mentoring. The websites of some representative universities that have well-established programs in technical, scientific, and professional communication were also reviewed for information regarding mentoring and internship programs. Lastly, selected Society for Technical Communication (STC) chapter websites were reviewed for information regarding mentoring programs. Information collected from these sources was used to provide a foundation for the analysis of mentoring models that includes survey information and the mentoring guidelines that are covered later in the chapter. These mentoring guidelines can also be useful to internship supervisors who often function as mentors to student interns. The information from these sources also provides the framework for the discussion of internships that follows.

REVIEW OF LITERATURE

Articles and books on mentoring focus on a range of mentoring models and activities. More has been written on the role that mentoring plays in industry, in management, and in psychology journals (Allen, Eby, Poteet, & Lentz, 2004; Eby & Allen, 2002; Johnson, 2002) than in journals devoted to technical and professional communication. For example, Branch (1999) notes that the presence of a formal corporate mentoring program is now viewed as one criterion by which the "Best Companies to Work For" are judged. However, in terms of technical and professional communication, Zimmerman and Paul (2007) examine the mentoring practices of teachers within the technical communication classroom, particularly the project-based technical communication classroom. One of their conclusions is that most technical writing programs lack a clear model for mentoring, because established business models for mentoring don't align well with academic models for research and teaching. Stowers and Barker (2010) examine the uses of coaching and mentoring as they apply to organizational communication professors. Spivey and Voss (2005) discuss a specific mentoring program involving professionals from the STC Orlando Central Florida chapter and graduating seniors in technical communication at the University of Central Florida. Their article also briefly reviews similar mentoring programs at a variety of local STC chapters and concludes that such programs successfully link the academy with industry.

The published literature on internships in technical, scientific, and professional communication is more extensive than the published literature on mentorships. The value of internships from the viewpoint of industry professionals and industry supervisors is well documented (Galtens, 2000; Roever, 2000; Sapp & Zhang, 2009). On a different note, Munger (2006) discusses strategies for managing internships as well as internship benefits. Sides and Mrvica (2007) is the most comprehensive study of internship partnerships within the field of communications media. In addition to providing a brief historical overview of the origin of internships tracing its theoretical underpinnings to Aristotle, Sides and Mrvica cover different internship models, various administrative issues, and different methods of assessment. They also discuss the advantages and disadvantages of required vs. optional internships and paid vs. unpaid internships. Their discussion draws on specifics from 25 years of internship experience in the Department of Communications Media at Fitchburg State College (Massachusetts) where the internship is a required, semester-long, 40-hour-a-week experience for all seniors. A more recent extensive study of internships in technical communication is Savage and Seible (2010). This study provides quantitative data on such topics as internship requirements, academic credits, and workplace hours for academic credit for over 70 programs.

Other publications in technical, scientific, and professional communication provide a range of perspectives and present a variety of practices relating to the topic. For example, drawing on the results of a survey of program directors, industry supervisors, and interns in technical communication programs, Little (1993) examines the use of internships in a variety of technical, scientific, and professional communication programs. Little's article also looks at internships through the conceptual framework of Kolb's learning-cycle model and concludes that internships are a valuable form of experiential learning. St.Amant (2003) proposes a stakeholder approach to the internship experience in technical communication. This approach argues for setting up colloquia involving all three stakeholders (the intern, the internship provider, and the academic advisor) to improve the quality of the internship experience. The stakeholder approach also creates student publication groups that involve students in work-world projects while they are being supervised by an academic advisor. Henze (2006) discusses the research-experiential internship and provides examples of student projects that have features of the research-experiential internship.

In examining the various aspects of internships, Tovey (2001) provides a detailed description of the 140-hour, three-credit internship program at East Carolina University (Greenville, North Carolina). Tovey also examines issues of socialization and acculturation facing interns in a workplace environment and concludes that internship programs can be a win/win situation for both industry and academia. Hart and Glick-Smith (1994), by contrast, focus on the broad connections between training and education and how these connections can result in better collaboration between the academia and industry.

ORIGINS AND NATURE OF MENTORING

Mentoring is one effective way that technical, scientific, and professional communication programs can collaborate with industry. While there is no universally accepted definition of the term "mentor," in classical Greek mythology, Mentor was an Ithacan noble in Homer's *Odyssey* and a friend of Odysseus. Mentor was entrusted with the care and protection of Telemachus, Odysseus' son, when Odysseus left for the Trojan War. Mentoring is very common in industry and most often associated with the career development of new employees.

The definition of mentoring used in this chapter is similar to the one found in the STC *Guidelines for Mentoring Programs*. That definition states that mentoring is a way for more experienced technical communicators to share their experiences with new or less experienced protégés (STC, 2002). For these reasons, it is important to research and review how different models of mentoring might be facilitating academy-industry relationships and partnerships.

Different Mentoring Models with Survey Results

The two models discussed in this chapter are the more formal mentoring model used by several STC chapters and the informal model that we use in the Department of English, Technical Communication, and Media Arts at Southern Polytechnic State University (Marietta, Georgia). These models provide the foundation for the observations made in this chapter, including the guidelines for building effective mentoring programs discussed later in this chapter. Both models successfully pair students with professionals in various communication fields to guide the student in his or her professional development. Both models view mentoring as a process that provides both the mentor and mentee with distinct benefits.

Mentoring Programs Sponsored by Representative STC Chapters

The STC Orlando Central Florida chapter mentoring program is a good example of the type of STC chapter mentoring models that exist. The STC Orlando Central Florida chapter has partnered with the technical communication program at the University of Central Florida to provide seniors, graduate students, or recent graduates with a mentor who provides career guidance (Spivey & Voss, 2005). The mentoring relationship lasts for at least 1 year and not more than 2 years. The mentorship program includes three positions: program coordinator, mentor, and mentee. The responsibilities of each position are described in detail. For example, the program coordinator assigns mentor and mentee pairings based on application forms that each potential mentor and mentee fill out. Other activities involving the program coordinator include serving as a resource for both parties and monitoring mentor/mentee activities by setting up a Facebook page expressly for this purpose.

The mentor's main responsibilities are to serve for two semesters or terms, to work with the mentee to define the objectives of the relationship, and to identify activities that will achieve the objectives. In addition, the mentor meets with the mentee once a month, provides feedback on the mentee's résumé and portfolio, and helps the mentee develop a networking plan. Finally, the mentor provides feedback about the mentoring experience to the program coordinator.

The mentee must participate fully in all of the mentor activities mentioned above that involve the mentee. In addition, the mentee must join STC and also join the local student STC chapter, be receptive to the constructive criticism of the mentor, provide feedback about the mentoring experience to the program coordinator, and make regular postings to the mentorship Facebook site.

The mentoring program set up by the STC Orlando Central Florida chapter and the STC student chapter at the University of Central Florida requires that both the mentor and mentee keep a contact log, submit two reports during the year, and participate in an assessment of the program at the end of the mentoring experience. The collaboration between the STC Orlando Central Florida chapter and the STC chapter at the University of Central Florida also provides a number of benefits for both the mentee and the mentor.

Mentees receive career guidance, develop networking skills, are exposed to different elements of a corporate culture, and receive feedback on a number of professional, career-related documents that they have written, including their portfolios. The mentor's benefits include helping a beginning professional get started in his or her career, learning about the curriculum of an academic program, and developing leadership skills. Links to a variety of forms used by the STC Orlando Central Florida's Mentoring Program (2011) can be downloaded at the Orlando chapter STC Education Mentoring Program website. These forms include guidelines and procedures, application forms, and an agreement form. There is also a link to the helpful *Invest in the Future . . . Sponsor an STC Student Mentoring Program* Reference Guide (Voss & Bowles, 2012).

Other STC chapters offer similar mentoring programs, although not a lot of information is available on the Internet. STC Toronto (2010) uses a 4-phase mentoring cycle that consists of preparing, negotiating, enabling, and coming to closure. This cycle is identical to the one Zachary (2000) describes. Mentors in the STC Toronto mentorship program must be senior STC members and be willing to commit 4 hours a month to the mentorship program. They also must complete a questionnaire sent to them by the mentorship program coordinator. Unlike the STC Orlando Central Florida's mentorship program, the STC Toronto program does not partner with any specific university or STC student chapter. Instead, the mentoring program is open to any STC member, although students in technical communication programs are encouraged to participate.

The Willamette Valley STC Chapter (2011) in Portland, Oregon, also has a mentoring program. Based on information on the chapter's website, the mentoring program is not specifically linked to a university or college student STC chapter. Instead, the Willamette chapter's mentoring program pairs new technical communicators with senior members to discuss career-related issues and technical problems. The mentor's responsibilities include listening, coaching, and guiding new writers. Other STC chapters that have mentoring programs are the Silicon Valley chapter; the Carolina chapter; the Rochester, New York, chapter; and the Lone Star chapter.

Taken together, these various STC mentoring programs are effective mechanisms for fostering academy-industry relationships. Mentees benefit enormously by receiving career guidance, networking opportunities, and feedback on a variety of career-related documents. Mentor benefits include the satisfaction in helping a new professional get started in his or her career and also the opportunity to keep current on academic programs. Programs looking to establish a mentoring component may want to borrow some of the techniques used by these mentoring programs described above. Doing so can help such programs create unique opportunities for fostering or maintaining academic-industry relationships and partnerships.

Unfortunately, a recent change in the tax law may alter the types of formal mentoring programs described above. As Hillary Hart noted in her 2011 Council for Programs in Technical and Scientific Communication keynote presentation, the IRS has apparently tightened the regulations on nonprofits, notifying STC that several student chapters were not in compliance. How the change in the long run will impact the existence of student chapters remains to be seen. For the short term, however, the change may threaten the existence of formal mentoring programs that involve pairing an STC chapter with a student STC chapter.

Informal Mentoring Programs

What may result is the growth of informal mentoring partnerships such as the successful one we have at Southern Polytechnic State University (SPSU). We began our mentorship program 10 years ago. Like other mentorship programs, ours is designed for seniors. All seniors in our Bachelor of Science in Technical Communication, Bachelor of Arts in English, and Professional Communication, and our Bachelor of Arts in New Media Arts are required to take our capstone course in Project Portfolio.

Unlike portfolio courses in other universities, we have included a mentorship program to help students bridge their academic careers with their new professional careers. We felt that the portfolio course was a logical place for a mentorship program to be included. I team-teach the course with Kim Haimes-Korn, who directs our writing program. Since the philosophy and goals of the portfolio course are instrumental for understanding our informal mentoring program, I have included a description of the course here:

> Project Portfolio is a 3-hour course that examines portfolios as professional tools for technical and professional communicators. In many ways, the portfolio is an ideal sales kit for technical and professional communicators because it demonstrates your accomplishments and skills through projects that you have already created.
>
> This course, which is team taught, will be conducted as discussion, workshop, lab work, presentation, and lecture. The course includes portfolio and

> writing theory within a collaborative workshop environment. Each student develops a professional portfolio of sample documents based on course projects, internships experiences, and/or work history. Essentially, you will revise these classroom assignments for portfolio quality and professional presentation. Since this course acts as a bridge between your time as a student and as a professional, we will discuss career options, interviewing techniques, résumé writing, and the job search process. At the end of the term, you will reflect (in writing) on your experiences as a student and as an entering technical or professional communicator.

Because the course covers career options, interviewing techniques, résumé writing, and the job search process, we pair each student with a professional communicator who mentors the student with career advice and also provides feedback on that student's developing portfolios. Unlike some of the more formal mentoring programs, mentors need not be STC members, although many of them are. Mentors also commit to a 1-semester time frame, the duration of the course, rather than a year or longer, as is the case with some formal mentoring programs. Often, however, students remain in touch with their mentors after the students have graduated.

The mentorship program coordinator, in our case, is one of the teachers of this course. Mentors are contacted well before the semester begins. Many mentors are alumni of our programs and/or are members of the local chapter of the Society for Technical Communication. Often they are web developers, technical writers and editors, graphic designers, marketing communicators, proposal writers, or work in social media. Because students in our Bachelor of Science and Bachelor of Arts programs pursue a concentration as part of their degree requirements, the coordinator of the mentorship program tries to pair each student with a mentor based on that student's concentration. During the first class meeting, each student mentions his or her concentration as part of a short introduction. Students concentrate in either information design, digital media and graphics, professional writing and communication, or new media. Students are also told that mentor pairings based on the student's concentration may not be possible and that they need to be realistic and flexible as to their expectations.

We have set up the mentoring program to involve the mentors and mentees in at least three activities. The program provides a certain amount of flexibility so that mentors and mentees can pursue activities that best meet the goals they identify during their initial face-to-face meeting. We decided to require an initial face-to-face meeting because each mentor wanted to meet the mentee in person. Often this initial meeting will take place at the mentor's place of work, where the mentee gets a chance to see the mentor's work environment and meet the mentor's coworkers. All of the other mentoring activities can be done through electronic mentoring or over the phone. These activities include

- Feedback on the mentee's résumé
- A review of the mentee's portfolio design
- A review of the mentee's finished electronic portfolio
- A review of the mentee's paper portfolio

While there is no formal agreement that a mentor and mentee sign, each mentee and mentor is expected to come up with a short list of mutually agreed-upon activities and a schedule of when correspondence will take place, hopefully, during the initial face-to-face meeting. Mentees provide regular updates on mentoring activities in class during the semester. We also involve mentors in a mentorship panel during a regularly scheduled class. This panel, usually consisting of three or four mentors, provides the class with career advice and also tips on how to create effective paper and electronic portfolios.

The SPSU mentorship program has produced some interesting opportunities for mentees. For example, one mentee had one of her portfolio pieces published in a company newsletter that featured a short article on SPSU's mentorship program. Needless to say, the mentee was very grateful for the exposure and recognition. The mentor of this student works for a large international electronics and electrical engineering company. The mentor's supervisor asked the mentor to write the article because this supervisor believed strongly in promoting mentorship programs, an attitude shared by many in industry. Other students received contract positions that grew out of the relationship between a mentor and mentee.

Based on feedback from the initial face-to-face interviews that mentees had with their mentors, the mentoring experience from the mentee's perspective was overwhelmingly positive. Jessica, whose mentor had a degree in technical communication from SPSU, but who now owns his own company that provides software services to companies, commented on the tips that her mentor provided her. She remarked that she felt more confident about the portfolio process and about employment opportunities in technical communication that would be available to her once she graduated. She also noted that her mentor provided her with valuable tips on how to market her degree to employers. For another mentee, the greatest value of the mentoring experience was that it underscored the importance of the concepts that she was learning in her classes. Her mentor, a Standard Operations Procedures Specialist for a global building-supplies corporation, identified that the most important skill that a technical communicator can have is the ability to do a thorough audience analysis. This link between the classroom and the workplace was an important one for the student.

In other write-ups of face-to-face interviews with mentors, mentees noted what they learned about the day-to-day activities of professional communicators. Ron remarked that he learned a great deal about what a technical writer does on the job. His mentor worked as a technical writer preparing proposals and graphics for technical documents used by an engineering firm.

Mentor reactions to the SPSU mentorship program point to the mentoring experience as a two-way street. A total of 13 mentors were electronically surveyed; 9 responded and 2 of the 9 mentors did not complete the survey for the following reasons: The student assigned to one of these mentors dropped the portfolio course, and the second student, according to the mentor, did not complete any of the required mentoring activities beyond the initial face-to-face interview. As a result, survey findings are limited due to the small sample size and low percentage of responses (7 responses, 53%).

The survey was given to the mentors who participated in the SPSU mentorship program for fall semester of 2011 (Smith, 2011). The survey consisted of the following core questions:

1. What motivated you to become a mentor?
2. How would you describe your relationship with your mentee?
3. As best as you can recall, what were some of the suggestions that you made to your mentee regarding career advice or regarding the mentee's portfolio?
4. Mentoring is often described as a relationship that benefits both the mentor and the mentee. What benefits, if any, did you receive from this experience?
5. What suggestions do you have for improving the mentoring program?

Based on survey findings, mentors were motivated to participate in the mentorship program for a number of different reasons. A common thread was a desire to give back. The survey also revealed that a key aspect of mentoring is the importance that mentors place on providing guidance and sharing their experiences with new professionals. As a result, individuals interested in offering effective mentoring programs might consider pairing students one on one with an experienced professional communicator who can offer this type of guidance.

When mentors described their relationship with their mentees, their descriptions varied. Many commented on the friendliness and attentiveness of the mentee, but they also remarked that some mentees were not fully engaged in the mentoring program. As one mentor noted, the mentee seemed to be doing it only because it was an assignment. The most productive mentor/mentee relationships occurred when rapport was established early on in the relationship and when both parties communicated frequently to develop trust and rapport. Consequently, individuals interested in creating mentoring programs may want to create programs that run longer than a semester, such as the STC Orlando Central Florida chapter's mentoring program, to maximize the potential for a mentoring relationship built on rapport and trust. When setting up a mentoring program, program directors may want to create a separate grade for the mentoring program and invite mentors to provide feedback on the grade.

When commenting on the types of suggestions they made to their mentees, some mentors focused on interview tips. One mentor suggested that the student be prepared to show examples of more than just school projects when possible.

These examples might be from volunteer work or internships. Other mentors focused on the importance of having students find a job that stirs passion and interest because those are the jobs that will motivate them the longest and ultimately produce their best work. Other suggestions spoke to the need for students to include only their best projects, combining academic projects with internship and volunteer projects. As a result, mentoring programs often work best when students are close to completing their programs and are working on capstone projects such as portfolios or internships. Internship supervisors can often provide a valuable service as mentors while internships are taking place.

The survey asked mentors if they benefited from the experience. All mentors noted that mentoring was a two-way street and that they benefited from the experience. One commented that an individual learns the most when in a position to teach others. Another mentor noted that the mentoring experience exposed her to new tools and technologies.

These responses and others like them also point to the dynamic nature of the mentoring process. Mentors felt that working with students helped mentors keep in touch with current technologies and also permitted mentors the opportunity to provide students with information about career opportunities. These observations point to the long-term value that mentoring programs play in building important relationships between mentee and mentor and academia and industry.

Finally, I asked the mentors for suggestions to improve the mentoring program. The responses were not as detailed as answers to some of the other questions and in that sense, were a little disappointing. One mentor suggested that we do a better job of preparing the student for the initial face-to-face meeting. Another suggested an initial event, perhaps an in-class panel discussion, which would involve all the mentors and students in a discussion of career options in technical communication. We hope to implement this suggestion in some form at a later date. Other mentors noted a preference for having the mentor and mentee meet in person more than once during the mentoring program. This suggestion is particularly useful if distance and schedules are not major obstacles preventing the mentor and mentee from meeting in person.

SOME GUIDELINES FOR BUILDING EFFECTIVE MENTORING PROGRAMS

Whether the mentoring program is a formal or informal arrangement, following certain guidelines helps make the experience a productive one for both the mentor and the mentee. The following guidelines are based on the previously reported research findings on mentoring:

- It is important for the mentor and mentee to establish rapport and trust early on in the relationship and to communicate frequently to cultivate that rapport and trust.

- The mentor and mentee need to set modest goals in writing; goals that can be realistically accomplished during the timeframe of the mentorship. Setting these goals is best done during the initial face-to-face meeting.
- The mentor and mentee need to develop and agree upon a schedule for achieving these goals.
- The mentoring relationship involving mentor and mentee should last at least one semester or one school term.
- The mentor and mentee must agree to remain accessible to each other so that frequent contact can occur.
- It is helpful if the mentorship coordinator can pair the mentee with an experienced professional who shares similar career interests.

Likewise, there are certain factors that will undercut the mentoring program and make it unsuccessful:

- The mentor and mentee are not a good match.
- Either the mentor or mentee has unrealistic goals for the mentorship program.
- Trust and rapport are never established.

In addition, the culture of the academy often does not encourage or support mentoring programs. For the most part, mentoring is usually treated as an additional faculty service load with little or no recognition and compensation. Nevertheless, as comments from students and mentors suggest, mentorship programs are very beneficial experiences for all involved and are an effective means for building strong academy-industry relationships and partnerships.

STUDENT INTERNSHIPS AS GATEWAY EXPERIENCES

Internships, like mentoring programs, can play a vital role in building strong ties between the academy and industry. As gateway experiences, internships are a valuable form of experiential learning, providing students with the opportunity to experience their chosen profession before they enter it full time. Internships also provide students with the opportunity to apply what they have learned in the classroom as well as learning something about how corporations operate and how they work with clients. Internship supervisors can also play an important mentoring role in coaching, guiding, and listening to students who are interning.

The origins of internships can be traced back to the trade guilds of the 11th and 12th centuries. Master tradesmen often brought on young, inexperienced, apprentices who worked under the guidance of one master tradesman to learn a particular trade by performing a variety of tasks associated with that trade. Apprentices most often worked with one master tradesman throughout their teenage years before they graduated to journeyman, earning better wages. Industrialization and formal education replaced the guild systems, and fields like medicine included the internship as a formal part of the education process. During the late 1970s, internships gradually became a popular form of experiential learning in other disciplines, particularly programs in business schools. Since then, internships have evolved into what we know them to be today—an important part of the formal education of many college students.

Interns in technical communication are often involved in a number of different types of activities. Some of the more representative activities are listed here.

- Preparing proposals
- Writing articles for publication
- Designing software documentation
- Writing blog posts
- Designing graphics
- Creating web pages
- Writing press releases
- Performing market research
- Writing reports
- Creating social media pieces
- Creating videos and multimedia pieces

Through such activities, students learn a great deal about the types of tasks that they may be doing professionally, in addition to learning the skills needed to do those tasks. Furthermore, students get the opportunity to network with working professionals and to be mentored by a supervisor who has professional experience. Supervisors who mentor may benefit from the effective mentoring practices presented earlier in this chapter (see the "Different Mentoring Models with Survey Results" section).

DIFFERENT INTERNSHIP PROGRAM MODELS

Internship programs are structured differently, in part, based on the importance placed on the internship within a specific academic program. Traditionally, internships involve three parties: the student intern, the provider of the internship, and the academic advisor (St.Amant, 2003). The character of the internship, however, depends on a number of variables that each technical communication program and university agree to use to set up the internship. These variables are listed here:

- Required or optional
- Credit or noncredit
- Length of the internship
- Paid or unpaid

Required or Optional

One of the more established internship programs is the one in the Department of Communications Media at Fitchburg State College. The internship program has been in existence since the mid-1970s and is a capstone graduation requirement for all students. The internship is a full-time, semester-long internship worth 12 credits out of a possible 120 credits (Sides & Mrvica, 2007). Because the student is interning full time, the student does not take other courses during that semester. In an internship survey that included 39 program directors from a variety of undergraduate programs in technical, scientific, and professional communication, Little (1993) notes that the percentage of programs requiring internships versus the percentage that make the internship optional is almost identical (44% requiring it and 51% not requiring it). In a later study, Savage and Seible (2010) note that fewer than half of the technical communication programs in their study required internships. Internships in the English, Technical Communication, and Media Arts programs at Southern Polytechnic State University are optional largely due to the number of students who work full-time or part-time.

Credit or Noncredit

As noted earlier, the internship program at Fitchburg State provides students with 12 semester credits since it is a full-time, semester-long internship. The vast majority of internship programs in Little's (1993) survey provided some form of academic credit for internships, although the number of credits varied from 1 to 9 units of credit. Little also notes that 41% of the programs offering internships require 180 hours or less of internship hours worked. Savage and Seible (2010) note that almost all the technical communication programs offering internships provide internship credit toward graduation, although the number of work hours required per credit hour varied widely. For-credit internships emphasize the important link between what is learned in the classroom and what is experienced in the workplace.

The internship program in the English, Technical Communication, and Media Arts Department at Southern Polytechnic State University requires 120 hours, usually 10–20 hours a week, for 3 hours of academic credit. This type of arrangement has worked well for us since it provides the student with an in-depth internship experience and is flexible enough to allow the student to take additional coursework and/or continue working on a part-time job. One of the conclusions, however, that Savage and Seible (2010) reach is that a 3-credit internship should require about 150 hours of on-site work. We may rethink our internship at Southern Polytechnic State University in the near future and require 150 hours.

Paid vs. Unpaid Internships

Programs seem to be almost equally divided as to whether they offer paid or unpaid internships (Little, 1993). Many offer both paid and unpaid internships. Arguments vary as to whether or not internships should be paid or unpaid. Those arguing for unpaid internships often mention that unpaid interns are given more professional responsibilities and end up having a better professional experience (Sides & Mrvica, 2007). Paid internships, however, tend to replicate the workplace experience more closely and also help to minimize the financial hardships on the student. It's been my experience that companies that provide paid internships seem to have a bigger investment in the student intern, which often results in more meaningful tasks. We provide both paid and unpaid internships. Often companies that contact our department provide paid internships. Students can also find their own internships, which are either unpaid or paid. If the student finds an internship, the department internship supervisor must approve the internship. To avoid potential problems, internship job descriptions from client companies must be very thorough.

Internship Qualifications and the Internship Evaluation

Most internship programs require the student to have some related academic coursework in the major before the student applies for an internship (Little, 1993). There is, however, no universal guideline as to how many hours in major the student should have before beginning an internship. Requiring the student to complete a substantial percentage of the coursework in a major seems to be a good guideline to follow because companies offering internships generally prefer to work with students who have completed a good part of their coursework. In some programs, the internship is a capstone graduation requirement (Sides & Mrvica, 2007). In other programs, like the one at Southern Polytechnic State, students need to complete a certain percentage of coursework in their major and maintain a certain grade point average. In our program, students need to complete at least 50% of the coursework in the major (most complete more) and also maintain a 3.0 average or better on a 4.0 scale in the major.

Understandably, other programs that require internships, particularly as capstone courses, have different requirements, sometimes based on the location of the internship. For example, the internship program at Fitchburg State College requires students to have a 2.5 on a 4.0 scale for an internship that is within a day's commute. If the internship is outside a day's commute, the student must have a 3.0 to qualify; if it is an international internship, the student needs a 3.3 (Sides & Mrvica, 2007). If the student decides to pursue an international internship, it might be wise to match that student with a mentor who can help the student prepare for cultural differences that might affect the internship experience.

Programs offering internships evaluate them very differently. Some will schedule visits to monitor the intern on the job, although this approach is not common due in part to the number of other departmental responsibilities an internship program supervisor may have. Interns, however, are commonly evaluated jointly by both the company supervisor and the internship program supervisor. Less clear is how much weight is given to the company supervisor's evaluation, usually an e-mail, memo, or filled-out printed form, in determining the intern's grade. To complete the internship, interns may submit a variety of documents, depending on the requirements of the program. In some cases, the intern will submit a portfolio, which is followed by an oral defense of the portfolio before a faculty committee (Sides & Mrvica, 2007). In other cases, the intern will turn in a final report, a reflective narrative, a journal of activities, and a revised résumé to include the internship (Sides & Mrvica, 2007). Our approach at Southern Polytechnic State University uses features from each of these approaches. Interns submit a final report, with artifacts completed during the internship that could be used as portfolio pieces. They also submit a reflective narrative, a weekly log of internship activities, and three progress reports. They also meet twice with the internship program supervisor. At these meetings, the internship program supervisor may mentor the intern if the intern is having any difficulties that involve the intern's supervisor at work. Finally, we recommend that the company supervisor submit a formal evaluation of the intern, although we don't require it. We may require it in the near future.

Supplementing the evaluation techniques discussed above is the stakeholder education approach, which includes two strategies. The first strategy focuses on creating regular industry-educator colloquia wherein educators, industry representatives, and interns meet at least once a year to discuss and improve internship practices. At that meeting, students who were interns reflect on the role that internships played in their education (St.Amant, 2003). The second strategy is the creation of student publication groups that complete real-world projects for corporations. While corporations submit the projects to a faculty advisor, the projects are supervised and completed by student publication groups (St.Amant, 2003). Through projects and programs such as these, one can greatly enhance or strengthen academy-industry relationships with partners within and beyond the college or university.

INTERNSHIP BENEFITS

The vast majority of students who intern find the experience very worthwhile. In her reflective narrative, one student remarked that the internship had given her valuable insights into what it would be like to work in public relations. The student noted that the internship experience helped her to clarify her career goals. Another noted that the internship gave him a good idea of what it would be like to have his own clients. Reflections like these are fairly common. Some of the more important benefits that interns and sponsoring companies/organizations are likely to receive from participating in internships are enumerated below.

For the Intern:

- Gain professional experience in major before graduation
- Apply knowledge and practices learned in the classroom to workplace problems and situations
- Bolster academic learning with experiential learning
- Create portfolio pieces
- Lessen educational expenses if the internship is paid
- Understand the expectations of a company or organization and the corporate/organizational culture
- Create networking opportunities
- Obtain recommendations
- Have the opportunity for a job after graduation
- Gain valuable experience that will be useful in the job search because employers often will not hire graduates who don't have some experience

For the Sponsoring Company/Organization:

- Develop long-term partnerships and research opportunities with academic programs
- Obtain knowledge about what is being taught in academic programs
- Contribute to the growth of the profession
- Benefit from the intern's up-to-date skills and knowledge
- Obtain extra help with minimal commitment
- Practice preemployment screening

APPLICATION OF IDEAS

This chapter has discussed the important roles that mentoring programs and internships play in the education of technical communication majors. These types of experiential learning activities can happen only when the academy and industry partner to promote the profession and jointly contribute to the student's total education. The chapter has also provided you with a description of formal and informal mentorship programs with guidelines on how you can set up each of these different mentorship programs.

Setting up a mentorship program can help academic programs build strong ties with industry. In addition, mentorship programs can provide mentees with valuable career guidance, networking opportunities, and feedback on a variety of professional, career-related documents. Sponsoring organizations benefit as well by providing guidance to beginning professionals and by staying current in what is being taught in academic programs. It is a win/win situation for both the academy and industry, and I recommend that a mentorship program be included in any capstone course or capstone experience for technical communication majors.

Likewise, internship programs augment what is taught in the classroom by providing students with valuable experiential learning. This chapter has provided you with a discussion of different internship models that can be easily modified to fit most, if not all, academic programs. This chapter also includes specific guidelines for setting up internships, establishing internship qualifications, and tips on how to evaluate internships, should your program not already include an internship component. Like mentorship programs, internships extend learning outside the classroom and provide both the intern and the sponsoring company/organization with specific benefits. As Sides and Mrvica (2007) note, internships can also promote a mentoring relationship between the supervisor at the company, the intern, and the intern supervisor so that the internship program continues to improve and grow.

The mentoring and internship experiences described in this chapter can foster or strengthen effective academy-industry partnerships and relationships in a number of ways. Mentoring programs and internships provide students the opportunity to apply what they learned in the classroom to workplace situations. Mentors can provide students with professional advice on a range of career decisions. Mentors can also draw on their industry experience to coach, guide, and provide students with feedback on a number of academic projects, including portfolios and internships.

CONCLUSION

Mentoring programs and internships are two effective ways that technical communication programs can extend learning beyond the classroom and build lasting partnerships with industry. Effectively managed mentorship programs provide both the mentee and mentor with distinct benefits. In short, it is a win/win situation for all involved. Mentees receive career guidance and feedback on a number of professional, career-related documents including their portfolios. Mentors get a chance to help beginning professionals get started in their careers as well as getting an opportunity to learn about the curriculum of an academic program.

Internships, like mentoring programs, can play a vital role in building strong ties between the academy and industry. As gateway experiences, internships are a valuable form of experiential learning, allowing students to experience their chosen profession before they enter it full time. Internships also reinforce the concepts and practices taught in the classroom in addition to exposing students to how corporations operate and work with clients.

This chapter has described different mentoring programs and internship models, noting the relative benefits of each. Mentoring programs such as the ones described in this chapter can be modified to fit many capstone courses, while internships can provide a very valuable stand-alone capstone experience. More research needs to be done on how mentoring can be included in technical communication programs. Likewise, more research is needed on how internships programs can be more effectively administered and how interns should be evaluated.

REFERENCES

Allen, T. D., Eby, L. T., Poteet, M. L., & Lentz, E. (2004). Career benefits associated with mentoring for protégés: Meta-analysis. *Journal of Applied Psychology, 89,* 127–136.

Bosley, D. (1995). Collaborative partnerships: Academia and industry working together. *Technical Communication, 42*(4), 611–619.

Branch, S. (1999). The 100 best companies to work for in America. *Fortune, 139,* 118–130.

Eby, L. T., & Allen, T. D. (2002). Further investigation of protégés' negative mentoring experiences. *Group & Organizational Management, 27*(4), 456–488.

Galtens, J. (2000). Lessons from the field: Socialization issues in writing and editing internships. *Business Communication Quarterly, 63*(1), 64–76.

Hart, H., & Glick-Smith, J. L. (1994). Training in technical communication: Ideas for a partnership between the academy and the workplace. *Technical Communication, 41*(3), 399–405.

Hayhoe, G. F. (1998). The academe-industry partnership: What's in it for all of us? *Technical Communication, 45*(1), 19–20.

Henze, B. R. (2006). The research-experiential internship in professional communication. *Technical Communication, 53*(3), 339–347.

Johnson, W. B. (2002). The intentional mentor: Strategies and guidelines for the practice of mentoring. *Professional Psychology: Research and Practice, 33*(1), 88–96.

Little, S. B. (1993). The technical communication internship: An application of experiential learning theory. *Journal of Business and Technical Communication, 7*(4), 423–451.

Munger, R. (2006). Participating in a technical communication internship. *Technical Communication, 53*(3), 326–338.

Roever, C. (2000). Mead Corporation's creative approach to internships: Success in a unionized manufacturing plant. *Business Communication Quarterly, 63*(1), 90–100.

Sapp, D., & Zhang, Q. (2009). Trends in industry supervisors' feedback on business communication internships. *Business Communication Quarterly, 72*(3), 274–288.

Savage, G. J., & Seible, M. K. (2010). Technical communication internship requirements in the academic economy: How we compare among ourselves and across other applied fields. *Journal of Technical Writing and Communication, 40*(1), 51–75.

Sides, C., & Mrvica, A. (2007). *Internships: Theory and practice.* Amityville, NY: Baywood.

Smith, H. (2011). *Project Portfolio Mentor Survey.* Retrieved October 18, 2012, from https://www.surveymonkey.com/MySurvey_EditorFull.aspx

Society for Technical Communication (STC). (2002). *Guidelines for mentoring programs.* AD-109-02. Available at AD-109-02.pdf.

Spivey, B., & Voss, D. (2005). *Bridging the gap between industry and academe.* 52nd annual STC Conference, Seattle, WA.

St.Amant, K. (2003). Expanding internships to enhance academic-industry relations: A perspective in stakeholder education. *Journal of Technical Writing and Communication, 33*(3), 231–241.

STC Orlando Chapter Education Committee. (2011). *Mentoring program.* Retrieved November 5, 2012, from http://www.stc-orlando.org/education/mentor/mentor.asp

STC Toronto. (2011). *Career day 2010.* Retrieved October 20, 2012, from http://www.stctoronto.org

STC Willamette Valley Chapter. (2011). *Mentoring program.* Retrieved November 5, 2012, from http://www.stcwvc.org/get_involved_mentoring_program.html

Stowers, R., & Barker, R. (2010). The coaching and mentoring process: The obvious knowledge and skill set for organizational communication professors. *Journal of Technical Writing and Communication, 40*(3), 363–371.

Tovey, J. (2001). Building connections between industry and university: Implementing an internship program at a regional university. *Technical Communication Quarterly, 10*(2), 225–239.

Voss, D., & Bowles, B. (2012). *Invest in the future . . . Sponsor an STC student mentoring program!* 59th annual STC international conference: The Summit, Rosemont, IL.

Zachary, L. J. (2000). *The mentor's guide: Facilitating effective learning relationships.* San Francisco, CA: Jossey-Boss.

Zimmerman, B. B., & Paul, D. (2007). Technical communication teachers as mentors in the classroom: Extending an invitation to students. *Technical Communication Quarterly, 16*(2), 175–200.

CHAPTER 6

Academic/Government Partnerships: Theoretical Underpinnings for Improving Online Writing

Barbara A. Heifferon

ABSTRACT

This chapter examines a partnership between a large State University and the State Department of Insurance. In this case, the Deputy Commissioner of the agency contracted with me about offering workshops to improve employee online writing skills. The project involved using certain rhetorical theories to persuade individuals to participate. The theories helped in the design and implementation of a persuasive frame for the workshops. The insights gained from this case can help professional communicators engage in such partnerships when teaching project management and design.

INTRODUCTION AND LAYOUT OF THE PROJECT

This academic/government partnership involved a university writing director and the State Department of Insurance. In 2011, the Deputy Commissioner of the department contracted with me to offer two workshops to improve employee online writing skills. This situation was related to the fact that many employees in this particular insurance office answered complicated insurance questions submitted via an online "Ask a Question" service. As soon as the director began laying out the parameters of this project, I realized the project would need rhetorical lenses to develop a persuasive frame for the participants.

In this chapter, I outline the project. I also explain the underpinnings of rhetoric and technical communication research that informed decisions about project design. I also discuss how the lessons learned from this project can be helpful in teaching undergraduates project design and implementation.

OVERVIEW OF THE CONTEXT OF THE PARTNERSHIP

In addition to overseeing the state's online "Ask a Question" operation, the recently hired Deputy Commissioner of the State Department of Insurance covers a number of areas: Management and Finance, which includes Administrative Services, Assessment and Data Management, Budget, Fiscal Affairs, Human Resources, Information Technology and Purchasing. Thus, training employees for the online operation seemed to combine Human Resources, Information Technology, and some management of data.

A former high school English teacher, the Deputy Commissioner was appalled not by the content knowledge of the employees, but by their online writing abilities. Their writing problems included stylistic choices that were too informal, grammar errors, spelling mistakes, and syntactical lapses. As a writing and technical communication professor, being asked to work only on grammar problems is not a focus I prefer to take up; however, given political realities in the state government, I chose to embrace this challenge. As Lutz and Fuller (2007) explain,

> One of the legacies of the traditionalist approach to business communication, for example, is a focus on genres, forms, and hard-core rules about grammar. And even postmodern teachers are ethically bound to teach social expectations about these features of language use. But professional writing teachers might need, in addition, to resist their urge to be corrective, to show that they are the experts. (p. 23)

I also remembered Miller's (1979) advice,

> Our teaching of writing should present mechanical skills and rules against a broader understanding of why and how to adjust or violate the rules, of the social implications of the roles a writer casts for himself or for herself and for the reader, and of the ethical repercussions of one's words. (p. 617)

Thus, in considering the narrowness of the tasks I was asked to complete and having been prepared as an academic in theoretical work, I saw I could design a broader and deeper understanding of writing for these employees within which to couch the need to communicate correctly.

As the organizer of this training session, a challenge I faced was addressing the skills that the Deputy Commissioner (D.C.) wanted improved while determining how to bring the employees into a more empowering conversation about audience and about their roles as writers. Our collaboration lasted 7 weeks and required several meetings to facilitate advance planning and development of a training program. The program itself would include two 3-hour workshops, homework assignments in between the workshops for the employees, and a follow-up visit for debriefing plus a deliverable. Before the workshops began, I met twice with the D.C. and her staff, reviewed the data they had provided, and planned the workshops. The two workshops were separated by a 3-week-long period employees used to complete assignments I gave them. The 7th week was used for debriefing with the D.C. and her staff.

A daunting constraint was that the D.C. did not want workshop participants to know they had been selected because of their poor writing abilities. As a good manager, she knew revealing an individual's participation in the workshop was related to their poor writing skills could adversely affect participation and could limit results. She also knew employees were challenged to communicate effectively with a public not likely to understand insurance jargon or government bureaucratic language. Employees also had to work using limited and limiting software. Like the study Popham (2003) conducted on the exchange of letters between health insurance systems and medical information systems,

> much of the communication between the medical activity systems and the business insurance systems results in contradictions of genres, objectives, and agency, and these contradictions are eventually negotiated through communicative means either in the medical practices or external to the medical practices. (p. 78)

In the Ask a Question system at this state agency, however, home and business insurance questions were more typical, although there were also some medical insurance queries.

These online insurance questions fielded by the employees were indeed rife with "contradictions" and varying "objectives" as Popham (2003) outlined previously. These, too, go beyond the surface-level errors, as do any attempts to address the perceived problem. While I could not turn the workshops at the Department of Insurance into graduate seminars on rhetoric, professional communication, or genre or activity system theory, it was important for me to understand "the complex relationship between context and text and the resulting social interaction" (Popham, 2003, p. 78). My hope was that I might help "the . . . insurance business [or government] community recognize their areas of interaction in ways that may improve their communications to make their online texts more effective, more efficient, and better suited to meet the needs of the [callers or clients]" (Popham, 2003, p. 78).

LITERATURE REVIEW AND THEORETICAL LENSES

The situation I sought to solve involved two problems: preventing workshop attendees from realizing they were selected because they represented the agency's "poorest" writers and deepening participant understanding of their roles as writers and the roles of their clients or callers in the writing/communication process.

A basic rhetorical understanding of audience was key to both items. As teachers, we always have this double task of teaching and being persuasive to our own audience (students or employees) as we then train or teach students/employees to consider their audiences. My first audience, however, was the employees, and my first task was to persuade them to take part in the project while not highlighting "deficiencies" in their online writing.

Unlike Popham's (2008) study of insurance letters, the department with which I was working demanded grammatically correct writing of employees responding to "Ask a Question" entries. Moreover, the employees would also concentrate on learning "strategies that teach students [or workplace writers] how to become experts themselves. Business writing teachers, for instance, could help students take responsibility for punctuation, grammar, and diction to reveal the rhythm and meaning of their language" (Lutz & Fuller, 2007, p. 23).

Identification and Consubstantiation

To accomplish this objective, I thought it would be helpful to bridge the gap between external professor (outsider) and employee (insider). Within this context, Burke's (1950) identification came to mind, along with Freire (1995) and Freire and Macedo's (1971) approach to pedagogy and Geertz (2000) and Foucault's (1980) "local knowledge." These ideas could be conjoined to formulate a lens "to find the available means of persuasion" (Aristotle, n.d.). These theorists are helpful when working with situated knowledge and on-site praxis. Because these five all addressed a close identification with others (in this case, those one teaches) and centered their theory on a thorough respect for their audience members, they could help me solve the problem of persuasion I had identified. Even better would be to call "local knowledge" *situated learning,* such as that described by Brent (2011), who said, "Situated learning suggests that highly context-dependent skills such as rhetorical performance are best learned—perhaps can only be learned—when learners are immersed in the real context in which such skills must be performed on a daily basis" (p. 400). In spite of the problematic constraints of a narrow task and writers chosen for their lack of ability, situatedness was a given we had in our favor because I knew I could design the writing workshops around the actual work done within the department.

Burke's (1950) notion of identification was key to the project, for I was aware of the strangeness or uniqueness (for the employees) of bringing in a writing teacher from the nearby university; it was an activity that had not been done there before. Thus, such teachers would be seen as outsiders or experts of sorts, with both the positives and negatives of that perspective. I would need a way to get the employees to see me as more like them than different so that they could identify with me. Thus, we could be sharing an agenda rather than having two different agendas.

Identification for Burke (1950) was also part of being "consubstantial" with someone or a group of people. "We are clearly in the region of rhetoric when considering the identifications whereby a specialized activity makes one a participant in some social or economic class. 'Belonging' in this sense is rhetorical" (p. 28). (In essence, "consubstantial" means being one with someone else. We share certain goals and commonalities as people. We are more alike than different from each other.) According to this perspective, to get buy-in from employees, I would need to find ways in which we all "belonged" to the same group vs. being seen as an outsider (a perspective that could easily result in resistance from the insiders).

One of the less egregious complaints regarding the writing of responses by employees was insurance-related messages containing allusions to a local football team or university's sports teams. (Such items were especially common as an extra line added under the signature of the worker: "Go [Mascots!]") This "error" would be difficult to explain, I thought, as I perused the error types I'd need to address in my workshop (see Table 1).

Using a local team's mascot was a matter of informal versus formal writing. Even if I explained that we had to pay attention to audience, that would fall on deaf ears in such an informal and sports-enthusiast culture. Actually, the correct assumption would be that the audience member feels the same way. Thus, this nod to a shared enthusiasm is in many ways not a problem at all, certainly not from an audience perspective. This "error" of including an allusion to sports would need to be addressed in terms of genres and the differences between formal and informal writing. Rather than seeing such informality as an error, this suggested change and instead should be positioned as a businesslike tone that was more appropriate for governmental discourse.

Additionally, I could explain that a businesslike tone highlights the writers' expertise. I knew I should appeal to the employees as Subject Matter Experts (SMEs) and try to find another identification that would make sense to the writers. I could (and did) note that citizens seeking help via the online system needed reassurance from the person answering the question. The client needed an expert, and expertise was better communicated in more formal rather than informal writing, as formal writing can enhance *ethos* (credibility) and give the online audience confidence in their knowledge (Diefenbach & Todnem, 2012, p. 43.)

True identification between the writers and the imported teacher could be enhanced using Freire and Macedo's (1995) pedagogy. Their work focuses on drawing on the expertise of students and showing that you respect them by acknowledging their abilities. Unfortunately, there were only a few weeks in

Table 1. Taxonomy of Errors (Condensed) with File Numbers and Quantity of Files

Error	File Number
Accept/Except	27761, 31698
Advise/advice	28190, 28914, 29067
A/An	28313, 2 more file numbers
Affected/effected	27664
It's/its	29126
Has/have	29130, 29131
That/which	28093, 29116, 29213, 29297, 28216
There/their/they're	31445
Active/passive voice shift	27727
Addresses questioner by first name	27622, 15 more files
Commas to separate items in series	27464, 29226
Comma following introductory clause	28205
Compound words or two words	27563, 28932
Dangling modifier	27271, 27273, 20 more files
Go *Mascots* Super Bowl Champs	27647, 7 more files
Possessive form error	28293
Pronoun "you(r)" used improperly	27483, 28997, 2 more files
Pronoun agreement with antecedent	27494, 14 more files
Run-on sentence	27440, 21 more files
Punctuation/possessive use	22864, 10 more files
Semicolon	27518, 28092
Spacing with commas	28904, 12 more files
Spelling/wrong word	27417, 112 more files
Subject/verb agreement	27464, 7 more files
Verb form (tense used improperly)	27704, 8 more files

which to enact this miracle of improved online writing. Freire's (1971) and Freire and Macedo's (1995) work could, however, help create a rhetorical situation in which students or employees could be motivated to learn. Yet using Freirean dialogic as only a tactic could be problematic: "In order to understand the meaning of dialogical practice, we have to put aside the simplistic understanding of dialogue as a mere technique" (Freire & Macedo, 1995, p. 379). Instead, Freire and Macedo characterized dialogue as an "epistemological relationship" (p. 379). In essence, the dialogue between the employees and me should create a shared knowledge.

Freire and Macedo (1995) also suggest we engage in dialogue not because we like (or don't like) one another, but because we recognize the social "process of knowing. . . . In this sense, dialogue presents itself as an indispensable component of the process of both learning and knowing" (p. 379). Of course, workshops should be dialogic; one would think that's the point, although in the first site visit (preworkshop), the D.C. and staff presented the genre of workshop as *lectures* about writing! Those of us who have taught writing reject that approach as ineffective. In a workshop, participants learn best by a hands-on process in which they are very much involved in the writing tasks rather than being passive recipients.

With Burkean (1950) and Freirean (1971, 1995) lenses, the task became determining how to reduce *barriers* and make the learning process truly dialogic. Thus, I needed not only to identify with the employees but have them identify with me. As I began mapping out the project, I referred to printouts the D.C. had given me. As my students' supervisor, the D.C. could review all the online responses provided by her employees, and she did so on a regular basis.

To help me coordinate and plan my workshop, the D.C. picked one month of online responses as representative and put them in a large box stacked with printouts of the online questions and answers these employees had conducted with their clients. One month was a good representative sample of the volume of work employees completed. Not only had the D.C. retrieved all these responses from the 20 people who would attend the workshops, but she and her staff had sorted and bundled them according to error type. As I reviewed these materials, I saw some egregious and some not-so-egregious ones, but what became noticeable was the expertise I observed in the documents.

Anyone who has attempted to read his or her health insurance payments and benefits can understand how arcane the jargon in such documents can be. Such items contain both the discursive practices of government and insurance and the legal and liability issues, which feature their own obfuscatory language. In reviewing sample responses from my prospective students, I noted that the employee mastery of the content of complicated insurance matters was impressive, and the ways to build dialogue and identification were clearer. Recognizing the employees as SMEs would be accurate and persuasive, given their prowess at answering insurance questions in detail. The knowledge they had developed was something I did not have, and they did. I could foreground their knowledge as most important, while beginning to build a dialogue with them about writing, rhetoric, and audience, including grammatical, stylistic, and syntactic errors.

To keep the instruction situated, the sentences and passages we wrote and examined would draw on this mastery. Using this approach meant the employees and I were "experts" in a sense, and we could come together to create a guide to improve online communication with the public. We belonged to a project group intent on the same goal. As Freire (1971) wrote, "Authentic education is not carried on as 'A' *for* 'B' or by 'A' *about* 'B,' but rather by 'A' *with* 'B,' mediated by the world—a world which impresses and challenges both parties, giving rise to views or opinions about it" (p. 74).

Local Knowledge

Local knowledge is essential to many situations wherein you are called upon to *correct* or *improve* something. Coming in as an outsider is what many of us do as professional communicators and as technical writing professors. We do this not only in our consulting work, collaboration in interdisciplinary academic projects, or in client-based graduate training, but also in our undergraduate classrooms. In these contexts, we must offer respect to those we seek to persuade to go along with our suggestions.

Generally, professional communicators understand the value of cultural studies and empirical research, and in both areas, we have turned to anthropology and appropriated some of its tools. Technical communicators, for example, value quantitative and qualitative studies. For ethnographies, we often turn to Geertz (2000) and other anthropologists, whose lenses have proven helpful in our field. Such is the case in Rose's (1996) "thick descriptions" in case studies of teaching, and the same is true for ethnographies done by Brice Heath (1999), Latour and Woolgar (1986), Brown and Dobrin (2004), and Dautermann (1997). Technical communicators often draw upon the ethnographic approach as we recognize that many of our service-learning sites, workplaces, and university classrooms are worthy of study in order to effectively analyze and introduce communicative interventions.

Cultural understandings were also important to my project design, but time was short. In Geertz's (2000) work, "To an ethnographer sorting through the machinery of distant ideas, the shapes of knowledge are always ineluctably local, indivisible from their instruments and encasements" (p. 4). Because my time frame was limited, I knew I could not obtain a thorough understanding of the scene. Ideally, a longer time frame could include shadowing various online writers, interviews, and perhaps think-aloud protocols. But my situation was not ideal. I would need to analyze the rhetorical situation in real time and with what little textual evidence I had (i.e., data sheets of online questions and responses, verbal input from the D.C. and staff, and visual taxonomizing of errors provided by the D.C.).

Yet my focus as participant/teacher/observer should also be the social value of the responses and the opportunity to see them not only as problematic in error but rich in local knowledge. Employees at the agency had demanding and challenging jobs. They were also supervised thoroughly (observation of all online responses, reminiscent of the panopticon). Any employee who could work successfully in this agency was fully capable of complex tasks using a depth and breadth of insider knowledge. Only with the ability to absorb, understand, and then communicate the complicated knowledge of the state's insurance industry, legal statutes, and local insurance companies could an employee continue in the job, as even our "worst" 20 writers had accomplished. What follows is an example of the "local knowledge" from the state Department of Insurance employees:

Question [from the online Ask a Question]:

I received my Farmers Ins Grp homeowners policy and it increased over 20% to $4,050/year starting Feb 2010. I need to look around and appreciate your help with any companies or advice you can recommend. Please forward this to other people who may be able to help me. And I know other people that need help also. This is for 1,925 sq ft Mxxxxxx home with Extended Replacement Cost; Choice Deluxe Coverage. A copy is attached. One [Hurricane] Gustav claim for $500 freezer loss. [Hurricane] Katrina was with Lloyds who paid about a $2,500 claim. And they increased Hurricane deductible from 2 to 3% or +$2,000 to $6,000. This is getting outrageous. They didn't even send Vaseline with the renewal package. Thank you very much for any help you can give.

Response:

Dear Xxxxxx:

In order for us to confirm that your rate is appropriate for your homeowner's policy through Farmers Insurance Group, you will be required to file a formal complaint through the following web address: xxxx. Regarding other potential insurance companies for homeowner's coverage, the Department of Insurance is unable to recommend any particular insurer or producer over another to purchase insurance coverage. As you stated in your question, shopping around and comparing policies and prices according to your needs and available insurers in your area is required. One way to shop for multiple insurers is through an independent agent in your area. Not all agents/producers are authorized to write for all insurers; each agent is appointed by insurers to write policies. By contacting as many agents as possible in your area, you will have more choices of insurers and policies in order to best suit your needs and budget.

Independent agents in your area may be found through the following web addresses: xxx and xxx. If I can assist be of further assistance, please contact me directly through xxx-xxx-xxxx.

Sincerely, XXXXX

Or this answer to another question:

Response:

Mr. XXXX. Reinstatement dates can not be backdated. The reinstatement date is the date that the reinstatement was processed. Your agency should not have submitted business between the dates 5/1/2010 – 5/18/2010. This also means that State Farm should not have accepted any business from your agency and can not pay you for business that they may have accepted during these dates. If you have any questions please feel free to contact me directly. The Commissioner of Insurance wishes to thank you for submitting your question. By asking us questions, we are better able to understand your needs and improve our services.

Sincerely,
XXXXX

In this interaction, we see that the first question is a complicated one regarding hurricane damage from two separate storms and multiple claims from two insurance companies. The employee replies with multiple sets of instructions for how the homeowner can file a claim and find less expensive insurance coverage for the future. In the other response to a business owner, we see a high level of detailed knowledge regarding an insurance company's policies. Thus, these answers demonstrate the specific "local knowledge" and the direction of clients to other sources of knowledge available to them.

In addition "local knowledge" in Geertz's (2000) work, highlights the "blurring of genres" (p. 19). He does not address all the blurring of *electronic* genres extant now, but for the writers selected for my workshops, *their* electronic genres have blurred. That is, many Americans learn e-mailing, online chatting, and texting with friends and family in personal and highly informal settings. Now these writers were asked to adopt the formal writing expected in a business or governmental setting while using what they considered to be a more informal medium. For example, while grammar and syntax are important in governmental discourse, in personal writing, they are not considered so important. Thus, in using electronic media to convey insurance-related information, some of the workshop participants blurred concepts of formality, genre expectations, and media.

Interestingly, most writers selected for my workshop were not young, and most had experienced formal writing in English classes before online writing was prevalent. Many did not have educations beyond an associate's degree, and those with degrees probably had taken a composition course, but that course could have involved more personal than professional writing. The key was to clarify that, in some settings (e.g., the Department) and for some audiences (e.g., the clients of Ask a Question), formal writing was required when communicating online.

In planning this workshop, I realized any writing correction would need to be contextualized within the institution itself. I also needed to be aware of an employee's status in the organization. Although I might recognize and appreciate their expertise and knowledge, supervisors in the agency might see only a lack of expertise or inadequate writing. Here Foucault's (1980) concept of *local knowledge* could be helpful.

Foucault (1980) defined local knowledge as a subjugated knowledge (i.e., knowledge that is made subservient to those holding more dominant positions in an organization). In short, subjugation was found in a hierarchical situation that outlined positions of power. Within this context, Foucault defined *local knowledge* as a "whole set of knowledges . . . disqualified as inadequate to their task or insufficiently elaborated: naive knowledges, located low down on the hierarchy, beneath the required level of cognition or scientificity" (p. 82).

Although Foucault (1980) wrote about medical patients, this concept of local knowledge applied to my employees/students. Because the agency heads felt a need to foreground the lack of writing skills, the "set of knowledges" (content expertise the employees *had* mastered) had been at least temporarily subjugated or discounted. By emphasizing a lack, the supervisors set up a situation in which an SME could be discounted. As a participant observer, I had to acknowledge my own discomfort with the tasks at hand: to somehow bring in grammar and writing lessons while aware of the problematic subject position of the employees. In essence, I had to achieve a double-voiced discourse to work with the manager while identifying with the workers I wanted to teach. As I interacted with the D.C. and other staff prior to the first workshop, I explained my project design without highlighting theory or culture. I highlighted the supervisor's need to protect and persuade those whom we had invited as "inadequate" writers.

ADDRESSING THE PROBLEM

I decided the best way to achieve the purpose of these workshops was to ask the employees to create a brief online handbook that could be used by the employees not at the workshops. The handbook would contain a style sheet and examples using employee knowledge and language about insurance questions with before-and-after, sentence-level corrections. By examples, I mean for every error type highlighted in the D.C.'s taxonomy, there would be a corresponding correction (see Tables 1 and 2).

Employees would complete the handbook as "homework" in the period between the two workshops. At the start of the first workshop, the writers and I addressed audience, as they knew the types of callers they worked with. I could highlight the idea workshop participants were chosen to create in this handbook due to their knowledge of the Ask a Question interface.

Through this approach, I could avoid offending participants in terms of the reason for which they were selected for the workshop. I realize that positioning participants as SMEs with construction of a handbook for *other* employees could be seen as manipulation or "flattery," such as that addressed by Socrates:

"In my opinion then, Gorgias, the whole of which rhetoric is a part is not an art at all, but the habit of a bold and ready wit, which knows how to manage mankind: this habit I sum up under the word 'flattery'" (Plato, 380 BCE.). However, as flattery is used here to motivate employees to improve their writing, it was worth such a charge, and the employees were indeed experts because of the intricate knowledge they possessed.

APPLICATION OF THEORETICAL DESIGN

As the first workshop began, I implemented the project design. We met in a computer lab with a projector and computer set up with some key points, hyperlinks, and examples of various online responses and questions. The attendees were all adults, representing a diverse group, many of whom had been working at the capitol in this capacity for many years. The group was composed of both sexes and several ethnicities, and attendees represented a mix of younger (near 30) and older (50- to 60-year-old) adults. They were perhaps skeptical and (some)even a bit frightened at first, but they soon relaxed and seemed less threatened.

After the introduction and setup of the workshop, we discussed their audiences and developed concrete pictures of those writing to the online Ask a Question. In these initial discussions, I addressed why we needed to understand our audience before we write to them online. Together we developed a multilayered view of their customers. By focusing their views on the audience, I helped employees "understand audience as a complex construct" (Lutz & Fuller, 2003, p. 23). The clients of Ask a Question included workers, businesspersons, students, and such. Employees knew experts usually did not use Ask a Question, but it was clear they thought everyone else who wrote into the agency did. The goal of Ask a Question was to give complete answers online to avoid the employees getting phone calls. Some employees had developed template/boilerplate answers, and individual employees were at times specialized in what they could and could not answer. Most, however, shared similar areas and answered a range of questions. Employees were aware their answers were checked by the D.C.

Knowledge Transfer and Activity Theory

When I mentioned the creation of a handbook, I circulated some handbooks I had brought with me because I knew it had been many years since some participants had been inside an English classroom, and memories of writing handbooks could be distant. Most students think of their writing experiences in schools as separate and not often applicable to the job unless they take a course in technical/business writing. In short, there is little knowledge transfer. Many scholars in our field do address this issue of knowledge transfer from the classroom to the workplace. For example, in "Bridging the Workplace and Academy," Blakeslee (2001) suggests contextual differences can create frustration, disorientation, and other difficulties when students move from classroom to workplace. The idea of the Department of Insurance employees writing a handbook also prompted similar concern and disorientation. The question became how to address such issues in order to offer effective training to attendees.

Other scholars have used activity theory to address the disorientation and difficulty of knowledge transfer (e.g., Engestrom, 1987; Russell, 1997; Spinuzzi, 1996). The issue often came down to employee/attendee perceptions of what they were to do. Simply put, the employees had no schema or paradigm to embrace the genre of an Ask a Question handbook; they were unfamiliar with the activity necessary to produce one. I therefore consulted activity theory.

Second-generation activity theory includes the exploration of reusable components: ones that could be applied to this lack of familiarity with the composition of a handbook. This theory suggested reusing questions and answers located in the Ask a Question database. I also considered the ideas of scholars such as Dobson, LeBlanc, and Burgoyne (2004), who based their second-generation activity work on the simple idea that designers consider the social and organizational contexts in which innovation was to be introduced. Dobson et al. further identify that

> this work begins on the basis that omission in this regard represents an overwhelming reason, in design, for poor matches with users' needs, misalignment with change policies and plans, confusions of roles and responsibilities in practice, and as a consequence, often very poor levels of technology uptake and use.

Their ideas became a heuristic I used for evaluating this project (e.g., what categories of information need to be considered). In essence, I combined Dobson et al.'s ideas with the listing of common error types provided by the D.C. (see Table 1) to create a framework that workshop attendees could use to create an effective handbook for the Ask a Question software. This heuristic was composed of the following five parts/questions:

- Validation of expected functionality: Does it do what was planned?
- Usability, particularly in creation and search: Can users do what they need?
- Differences among user's experiences with metadata: What variety of needs exist?
- Fit with activity workflow: How would it fit with current work practice?
- Potential for integration: What is the likelihood of adoption?

This activity-based approach allowed me to collaborate with other workshop attendees to develop a handbook that could function as a quick, online (or on desktop) reference. All participants felt handbook users could accomplish their goals (i.e., learn about not repeating the same error) provided they used the handbook.

In terms of addressing the item, "Differences among user's experiences with metadata: What variety of needs exist?" the attendees and I felt the taxonomy developed by the D.C. and her staff was enough to start addressing user needs. We also acknowledged that the handbook could function as a living document or template in which other error types could be incorporated over time. Given the work done in the two workshops, which both the D.C. and her chief staff member attended, the two supervisors could also duplicate and continue the process of adding to the handbook with the same or other groups of employees.

The fourth item in the heuristic ("Fit with activity workflow: How would it fit with current work practice?") was initially daunting. Upon conferring with the D.C., I learned that the employees had a number of icons with insurance statutes and other reference guides on their computer desktops. This factor meant that certain actions (consulting an icon representing the handbook on the same computer desktop) would duplicate other parts of the workflow and could insure knowledge transfer. For this reason, we (the attendees, supervisors, and I) decided that the handbook should be available on their work computers, with a shortcut to the desktop.

When confronted with the idea of making their own handbook, the employees pondered the parameters of their own tasks within our joint venture. There was concern about how much writing and homework employees would need to do for the online handbook, and I explained that these handbooks were to be brief, and we would use a collaborative writing process to address the items with which their fellow employees had problems. We then reviewed other examples of small, online handbooks to gain a picture of the task. Interestingly, the idea of a collaborative writing assignment to compose a "text" was foreign to attendees.

The next important step was to review what would go into the handbook entries for this project. Table 2 provides some typical examples of such entries.

To create a fast and efficient online handbook, I had already used categories of errors provided by the D.C. to set up short chapters. I then furnished print and online templates for the employees. As homework, each employee used these templates to write a brief chapter addressing certain stylistic problems or a group of errors. (See Table 3 for an example of the homework assignment distributed to employees.) Also, unknown to the employees was that the topics I assigned them to work on in their chapters were the same items the D.C. had identified as one of their individual writing weaknesses. Thus, individual workshop attendees had specific errors on which to work, and in the 3 weeks in between the two workshops, each person's homework involved developing examples of writing problems and creating guidelines for addressing that problem (as well as corrected examples of the related writing problem).

Table 2. Examples of Incorrect Verb Forms: Tenses, Run-Ons, Fragments

Examples of Incorrect Verb Forms:

Mr. Jason Atchison license has been cancel as of 2/11/10, a letter of clearance was process and mail to the address that he requested.

The reinstatement date is the date that your reinstatement was process.

Your address should be change to your current address.

Examples of Run-On Sentence or Sentence Fragment:

Homewise Insurance Company has been licensed to do insurance business in Xxxx since 4/29/2008, they have 4 Property and Casualty consumer complaints on file, they are located out of Tampa, FL.

However, it would be assumed that if contaminated water reached your property this would be flood issue that would not be covered under the homeowners policy anyway.

If you do not receive this form within 45 days. Please contact the U.S. Department of Labor at x-xxx-xxx-xxxx.

In assigning this task, I suggested attendees consult grammar handbooks at their workplace and suggested they consult an online grammar source. Interestingly, many were unaware of the wealth of online grammar checks available to them. For this reason, I had attendees review a number of those toward the end of the first workshop. I also introduced them to the Purdue University's Online Writing Lab (OWL), showed them how to get to the OWL grammar sections assigned to them, and explained the worksheets, which they could compose for their fellow workers as the appendix of the handbook. To familiarize attendees with these resources, I encouraged them to put a grammar problem and/or rule in their own words or to quote a grammar rule found in the public domain.

The D.C., staff, and I collected the first part of the homework and studied the chapters after 2 weeks of the 3-week homework period. For the remaining homework time, the participants developed worksheets for the back of the small handbook. In the second workshop, we collectively reviewed a number of the submitted chapters and assessed the effectiveness of a number of them. As the D.C. and I had studied these in advance, I reviewed the most egregious chapters (anonymously) with attendees to help them better understand the related error. The good rapport and trust established in the first workshop allowed us to

Table 3. Homework for Workshop 2 State Department of Insurance

You are helping develop a style sheet/mini handbook for the XXX Department of Insurance to be used by those who answer online queries regarding insurance problems. Each of you has received a type or category of error that you will address in your contribution to this guide. Here is what you will do with the error or errors you have been assigned to address:

Make a **Header** for your category, for example: **SPELLING ERRORS** in all capital letters, in bold, in Arial font and point size 12.

Problem: Describe in your own words what the problems are: Spelling errors are a problem in online answers to insurance questions and can cause confusion for the reader. Spelling errors can consist of double letters where there should be only one and one letter where there should be double letters. Consonants and vowels can be misused. Some words are so poorly spelled as to be unrecognizable.

Examples of misspellings used in sentences:

Did you communicate with your agent/producer or the company directly for an explanation for such a large decrease?

If there are hours left over form the prior renewal period, only classroom hours are eligible to carry forward.

Please feel free to email me any additional questions you might have.

Examples of corrected spellings used in sentences:

Did you communicate with your agent/producer or the company directly for an explanation for such a large decrease?

If there are hours left over from the prior renewal period, only classroom hours are eligible to carry forward.

Please feel free to email me any additional questions you might have.

Worksheets or exercises for practice: Provide several exercises/worksheets. One good source for these is the Purdue OWL (Internet). Attach several to your homework. You can also make up your own practice sheets. Be sure to give credit, if you used a source in your homework.

Solution: To avoid misspellings in the online answers you give to people who write in: Write your answers in MS Word first and then copy and paste it into the online space provided.

critique and revise these chapters with good humor, and the compliance among employees on homework was 95%.

Also, after reviewing several chapters, I assigned attendees three questions to work on in class and to compose aloud. This activity is often used in group composition, and for it, we discussed ways to articulate questions like, Is this answer going to be clear enough for the person needing this information? For this part of the workshop, we used typical questions from Ask a Question:

> I have been convicted of felony theft and have served 3 years and completed 3 years of parole. I would like to prove to myself as well as to society that I have become a better person and would like to make an honest living by selling insurance. I would like to know if with my record, I would be eligible to apply for a license to sell insurance. Please consider my request and allow me the opportunity to prove that I can be an honest citizen.
>
> I was recently laid off from my job. My boyfriend asked his health insurance company if I could be covered under his policy as we live together, and he had heard that some insurance policies cover domestic partners. His insurance company told him to ask the HR department of his company. They told him no, they would not cover me because it is against state law to cover a domestic partner. Is this true?
>
> With regard to the Affordable Care Act, does Medicare cover a mammogram every year over age 65, or is it less often unless one is high risk? We have called Medicare, but cannot get a clear answer. Also, we are unsure regarding the no copay/deductible for mammograms for private insurance companies. Can you clarify if this is for all private insurance companies?

The chapter reviews and the collaborative composition made up a good portion of the second workshop session. During the session, participants were eager to discuss how to write the sentences (I composed on screen while they dictated) and how to correct example chapters. Especially effective was to hear attendees explain why one choice was better than another, and I heard them using some of the terms and rhetorical or writing concepts introduced during the two workshops. For example, Nellie G. said,

> Jose, if we word it that way, it's going to be less clear to our audience. You have good expertise there, but you are not communicating your expertise or this difficult information to your audience. You want them to trust you and that sentence is kind of tangled up and you don't sound as smart as we know you are.

This project allowed me to analyze this process and to identify a number of practices that could be of help in both organizing undergraduate and graduate classes as well as forging academy-industry partnerships around writing workshops offered to local organizations. I've presented these practices in the following list:

- It's most useful for individuals to be involved in teaching and training others, for at some point in almost any career or job, individuals will be responsible for some type of training of one or more employees.
- Studying rhetoric and the art of persuasion along with pedagogical theorists and professional communication research is most useful for us and students. These were central to my notions of audience analysis, the necessity to persuade and how to persuade and what strategies to employ in order to accomplish the purpose of this contract. Many places and programs are beset by challenges to humanities and rhetoric programs, and we sometimes hear these challenges even within our own English or humanities departments. Giving examples of how theory informs practice and creates *praxis* is an old argument in our field, but it can be beneficially reprised.
- Cultural sensitivity can also be reprised. We as technical/professional communicators are usually well versed in understanding that cultures of workplaces are crucial. We can use more theoretical research, case studies, and ethnographies of workplaces to develop effective communication strategies.

LESSONS LEARNED FROM THIS CASE STUDY

I hope the case presented in this chapter provides an example for other academy-industry partnerships. Key items to consider when using activities such as workshops to forge such relationships include

- Interrogate partnership requests if they seem to require tasks that go against the best practices in our field.
- Use theory to interrogate and create alternative emphases within those requested tasks.
- Use theory to recast the rhetorical situation and discover the available means of persuasion should one be asked to address a problematic situation on behalf of an industry partner (e.g., creating workshops to address a particular communication need as perceived by that partner).
- Focus activities such as workshops on hands-on deliverables that the attendees (versus the workshop leader) complete.
- Respect participants for what they know when asked to develop solutions or approaches involving those participants.
- Act as mediator, liaison, or consensus builder between academic and industry partners.

By applying ideas such as these when entering into partnerships with industry, academics/educators can establish experiences and relationships that work best for all involved parties.

CONCLUSION

In some settings, we are asked to participate in partnership interactions wherein the fit between academy and industry participants might seem more natural. But in many government agencies, bureaucracies, and industries, the culture is indeed "foreign" for us. It is therefore important we find ways to bridge these workplaces so our work can affect the real world and solve real-world problems. Because we are "foreigners" to some extent in these environments, we need to proceed with creativity, calling on the backgrounds we have and making use of our theoretical approaches. Our students need to learn how to do the same as they move into industry or government. It was Dixie Goswami (personal interview, 2005), quoting Kurt Lewin (1951, p. 169), the German psychologist, who said, "There's nothing more practical than a good theory" (conversation).[1] Indeed, our theoretical understandings are where we go to analyze and solve problems. Our graduate and undergraduate students often protest learning theory, seeing software mastery as more important. They are wrong. We must show our theory in action to help students understand why we teach it. Because many of us believe in *praxis,* we need to develop examples for students and honor *praxis* in the workplaces we visit.

Students need to understand the value of culture. Experiences we provide in ethnographic/case study or qualitative empirical research are valuable. Students need to be in the field doing the work. The more practice we can give them, the more successful they will be in making the transition to industry or government (or even to tenure track). We also need to reach out to those professionals with whom we work, as Academia is a very foreign culture to most. Because many of us are in university settings, when we are there a while, we forget how different our culture appears and we need to build bridges by explaining why we think the way we do.

Lastly, we need to embrace our theories and not be afraid to combine them in what may seem eclectic ways. Being confronted with a problem to solve, my task initially seemed impossible. After my first site visit, I felt I faced a Herculean task. Because of our *oeuvre* of good theory work in professional communication and rhetoric, when given tough tasks, we need to grab our toolboxes and pull out tools/theories. Given that I was an outsider and already had been positioned as an "expert," "identification" to achieve a sense of belonging, honoring "local knowledge" and Freirean "dialogics" were most effective in this particular workplace.

[1] "The consistent theme in all Kurt Lewin's work, according to David A. Kolb (1984, p. 9) was his concern for the integration of theory and practice. This was symbolized in Lewin's best-known quotation: "There is nothing so practical as a good theory" (1951, p. 169). It's a lesson that we still need to learn.

REFERENCES

Aristotle. (n.d.). Rhetoric. Book I (W. R. Roberts, Trans.). *The Internet Classics Archive*. Boston, MIT Online. Retrieved October 20, 2012, from http://classics.mit.edu/Aristotle/rhetoric.mb.txt

Blakeslee, A. M. (2001) Bridging the workplace and academy. *Technical Communication Quarterly, 10*(2), 169–193.

Brent, D. (2011). Transfer, transformation, and rhetorical knowledge: Insights from transfer theory. *Journal of Business and Technical Communication, 25,* 396. Originally published online June 2011. doi:10.1177/1050651911410951. Retrieved October 20, 2012, from http://jbt.sagepub.com/content/25/4/396

Brown, S. G., & Dobrin, S. I. (2004). *Ethnography unbound: From theory shock to cultural praxis*. Albany, NY: SUNY Press.

Burke, K. (1989). *On symbols and society*. Chicago, IL: Chicago University Press.

Dautermann, J. (1997). *Writing at Good Hope*. Greenwich, CT: Ablex.

Diefenbach, T., & Todnem, R. (2012). *Reinventing hierarchy and bureaucracy: From the bureau to network organizations*. Bingley, UK: Emerald Group.

Dobson, M., LeBlanc, D., & Burgoyne, D. (2004). Transforming tensions in learning technology design operationalizing activity theory. *Canadian Journal of Learning and Technology, 30*(1). Retrieved October 20, 2012, from http://www.cjlt.ca/index.php/cjlt/article/view/ 116/110

Engestrom, Y. (1987). *Learning by expanding: An activity theoretical to developmental research*. Helsinki, Finland: Orienta-Konsultit Oy.

Flowers, L., Hayes, J. R., & Swarts, H. (1983, 1999). Revising functional documents: The scenario principle. In P. V. Anderson, R. J. Brockmann, & C. R. Miller (Eds.), *New essays in technical and scientific communication: Research, theory and practice* (pp. 41–58). Amityville, NY: Baywood.

Foucault, M. (1980). *Power/knowledge*. New York, NY: Pantheon.

Freire, P. (1971). *Pedagogy of the oppressed*. New York, NY: Herder.

Freire, P., & Macedo, D. (1995). A dialogue: Culture, language and race. *Harvard Educational Review, 65*(3), 377–402.

Geertz, C. (2000). *Local knowledge: Further essays in interpretive anthropology*. New York, NY: Basic.

Heath, S. B. (1999). *Ways with words*. Cambridge, MA: Cambridge University Press.

Kolbe, D. A. (1984). *Experiential learning. Experience as the source of learning and development*. Englewood Cliffs, NJ: Prentice-Hall.

Latour, B., & Woolgar, S. (1986). *Laboratory life: The construction of scientific facts*. Princeton, NJ: Princeton University Press.

Lewin, K. (1951). *Field theory in social science: Selected theoretical papers* (D. Cartwright, Ed.). New York, NY: Harper & Row.

Lukensmeyer, C. J., & Torres, L. H. (2006). Public deliberation: A manager's guide to citizen engagement. *IBM Center for the Business of Government Collaboration Series*. Retrieved November 5, 2012, from http://www.businessofgovernment.org/pdfs/Lukensmeyer Report.pdf

Lutz, J., & Fuller, M. (2007). Exploring authority: A case study of a composition and a professional writing classroom. *Technical Communication Quarterly, 16*(2), 201–232.

Miller, C. R. (1979). A humanistic rationale for technical writing. *College English, 40*(6), 610–617.

Plato (380 BCE). Gorgias (B. Jowett, Trans.). *The Internet Classics Archive.* Boston, MA: MIT Press. Retrieved October 30, 2012, from http://classics.mit.edu/Plato/gorgias.html

Popham, S. (2003). The intersections of medicine and business: Case studies of contradictions and negotiations in medical-business writing. *Business Communication Quarterly, 66*(2), 76–78.

Popham, S. (2008). As if good writing mattered: A rhetoric of pragmatism in health care business letters. In B. A. Heifferon & S. C. Brown (Eds.), *Rhetoric of healthcare: Essays toward a new disciplinary study*. New York, NY: Hampton.

Rose, M. (1996). *Possible lives: The promise of public education.* New York, NY: Penguin.

Russell, D. (1997). Rethinking genre in school and society: An activity theory analysis. *Written Communication, 14,* 504–554.

Spinuzzi, C. (1996). Pseudotransactionality, activity theory, and professional writing instruction. *Technical Communication Quarterly, 5,* 295–308.

SECTION III

Engagement Venues

CHAPTER 7

Engagement Through Emerging Technologies: A Humanistic Perspective on Academe-Industry Relationships and Partnerships

Russell G. Carpenter

ABSTRACT

This case study analyzes a partnership linking the Noel Studio for Academic Creativity at Eastern Kentucky University (EKU) with a middle school and the Bluegrass Army Depot (BGAD) in Richmond, Kentucky. I use a humanistic lens to explore the role of emerging technologies in facilitating partnerships and argue that this approach will be informative in the successful design of future partnerships, including those developed in rural areas and that there is value in focusing on the humanistic perspective as it centers partnerships on the participants and relationships developed.

INTRODUCTION

As Bowdon (2011) notes, any educator who has attempted to create a genuine partnership with an organization outside of the academy can attest to the complexity of establishing balanced, reciprocal relationships that benefit all parties. Just managing the logistics and politics of getting the appropriate parties to a shared table is challenging enough, but those steps are only the beginning. The importance of shared vision in academy-industry partnerships cannot be overstated. Partners from the academic side often advance literacy initiatives, while partners from the industry side bring professional-development opportunities for extending these initiatives into communities of professionals, youth, or citizens. These factors came to the forefront when the author's home institution, Eastern Kentucky University (EKU), attempted to embark on such a relationship with a local community partner.

In this chapter, I analyze the complex process of partnership formation involving EKU and the Bluegrass Army Depot (BGAD) as well as a number of related stakeholder groups. To do so, I use a humanistic lens that focuses on the relationships among the people and institutions involved by providing agency to participants in the partnership. This framing can be valuable across many academe-industry partnership designs because it focuses our attention on the people involved in these partnerships. Such a level of understanding, moreover, can be particularly important or beneficial in creating effective partnerships in rural areas or regions where access to technology and communication programs can be scarce.

REVIEW OF LITERATURE

The purpose of this chapter is to apply a humanistic perspective as a theoretical lens to examine technical communication partnerships between academe and industry. This section presents a review of the literature in academe-industry partnerships. It offers an overview of academe-industry relations; humanistic perspectives for technical communication partnerships; the role of service learning in forming partnerships; emerging technologies and technical communication programs; and challenges facing academe-industry partnerships.

An Overview of Academe-Industry Partnerships

A report from the Engineering Advisory Committee Subcommittee on Industry-University Partnerships (2008) suggests that one way to incentivize partnership efforts is to promote academe-industry collaborations (or collaboration opportunities) that are diverse enough to meet the needs of businesses and universities of all sizes. Although the conversation has gained momentum in recent years, it is by no means new. Culliton (1983) summarized a conference on academe-industry partnerships held at the University of Pennsylvania, reporting that conference participants were equally divided between individuals from the academy and persons from industry. The primary question at hand during the conference: Can academic researchers enter into contracts without sacrificing important university values (p. 150)? The results coming out of the conference included employing emerging technologies to develop academe-industry partnerships and preparing technical communication students for the professional workplace.

In reviewing this event, Culliton (1983) highlights the amount of funding industry dedicates to research and development in universities. For example, in 1983 when this conference took place, industry spent "somewhere between $250 million and $300 million a year on research and development in universities" (p. 150). The partnerships discussed by Culliton also foreground the areas of business, engineering, and biology by noting that they commonly seek industry support. With the prevalence of partnerships throughout the world, it is clear that new opportunities for analysis have emerged.

To examine this context, I situate this chapter within the many partnerships explored by technical communicators, some of which were featured in the 2011 collection *Higher Education, Emerging Technologies, and Community Partnerships* (edited by Bowdon and Carpenter). This recent scholarship has offered successful models, concepts, and practices for establishing partnerships, assembling key scholars from around the country, and internationally, to consider perspectives on these relationships.

A Humanistic Perspective for Partnerships

As academe-industry relationships take center stage in a growing number of technical communication programs, the literature on the resulting partnerships formed by these relations needs further development both in breadth and depth. Within this context, Miller (1979) advocates for a humanistic rationale that focuses on the teaching of technical communication, a role in which teaching helps members of the field understand how to belong to a community (p. 617). In discussing such ideas, Miller provides an appropriate starting point for analyzing academe-industry partnerships: that to engage in communication is to participate in a community (p. 617). That is, Miller introduces the social implications of technical writing.

A humanistic approach to partnership design can also reveal important perspectives for determining best practices, goals, and future directions. Furthermore, this approach also takes into account place-based needs and community expectations. As Miller (1979) argues, a humanistic perspective supports the notion that there is value in examining the activity. I, in turn, argue that the activity is the partnership itself—the bridge linking academe-industry or the content of the relationship.

Building on Miller's earlier work, Dombrowski (1994) argues that social constructions should take priority over the dualities of science versus the humanities or objectivity versus subjectivity. In academe-industry partnerships, technical communicators serve as members of the communities with which they engage. Dombrowski also provides an appropriate overview of a humanistic approach to technical communication that serves as an entry point into the arguments offered in this chapter:

> Broadly speaking, humanism is the emphasis of the human over the non-human. It involves studies turned more toward humankind itself than

> toward the physical, non-human world, for example, toward literature or ethics. This humanistic emphasis is reflected in technical communication studies as interest in persuasion, in psychological and sociological constructs, in human technological practices, and in gender-fairness—all of these emphasized over the objective, material, non-human things that are often taken as the basic subject of technical communication. Thus, from the perspective of humanism, the root subject of technical communication is always humankind. (p. 4)

While the concept that partnerships should consider humanistic perspectives may not be entirely new, it is one that requires emphasis as academe-industry relationships are developed. In practice, the partnership described in this chapter develops interests in technical communication among participants. As part of the experience, students explored the role of audience and community in their communication.

The Role of Service Learning in Forming Partnerships

Santoro and Betts (2002) suggest that the shift to a knowledge-based economy has brought about a shift in relationships from "sponsorships" to "partnerships." Within this framework, universities offer knowledge not available in industry. Their research supports the establishment of partnerships for productive purposes, including tangible outcomes such as patents, licenses, and products (p. 42). Academe-industry partnerships, in turn, offer students, faculty, and universities the opportunity to engage in service learning. McEachern (2001) offers perspective on service learning in technical communication, focusing on nonprofits. His work outlines some of the challenges involved in working with nonprofits, including the lack of funding, resources, and time. Partnerships with nonprofits are especially prevalent in service-learning initiatives, which are common within technical communication programs (Bowdon & Scott, 2003).

Tovey (2001) highlights additional benefits and challenges with establishing partnerships through internships at her university. She does so by focusing on a service-learning approach in which she highlights the potential benefit of partnerships to students and faculty as the establishment of new connections and visibility for technical communication programs. Similarly, DeBerg (2011) offers perspective on university, high school, and private-sector partnerships with four key stakeholders:

- A nonprofit organization called Students for the Advancement of Global Entrepreneurship (SAGE)
- High school students and their teachers
- University students

- The private sector, including corporations, foundations, and individual philanthropists (p. 16)

Although McEachern (2001) foregrounds pressing challenges involving non-profits of particular interest to technical communicators, Tovey (2001) suggests that factors such as place, space, and time inevitably enhance or hinder efforts between academe-industry partners.

The scope of such partnerships, moreover, can be reconsidered in terms of the kinds of access communication technologies can offer to other organizations. Detailing the results of a Council for Programs in Technical and Scientific Communication (CPTSC)-funded study of program partnerships, Starke-Meyerring, Duin, and Palvetzian (2007) overview literacies that technical communication professionals must develop for use in partnerships. These include networked learning environments and experiential learning opportunities. Importantly, Starke-Meyerring et al. highlight the environment—or cultures—of technical communication partnerships:

> As we show, these changes increasingly call for new program partnerships to facilitate learning environments that immerse TC students in global digital networks with professionals, peers, citizens, and experts from diverse contexts; challenge students to negotiate and build shared learning cultures across diverse boundaries; and provide students with new opportunities for civic engagement in a global context. However, these are not the only changes calling for new partnerships. These partnerships also play an increasingly important role for TC programs as they position themselves amid current globalization processes in higher education. (p. 140)

Building upon these ideas, universities, including those serving rural communities, might leverage communication technologies to engage industry and community partners across a wider geographic range in productive relationships with mutual benefits. Additionally, such technologies could become a factor around which future partnerships might be organized.

Emerging Technology, Partnership Design, and Technical Communication Programs

Technical communication scholars have explored methods for designing partnerships, and these include the use of emerging technologies to foster interactions. For example, Henry (2011) documents a hybrid face-to-face (F2F) and virtual collaboration between the State of Hawai'i's Division of Forestry and Wildlife and a team of his technical writing students at the University of Hawai'i. Others, such as Cates, Cedercreutz, Harfmann, Lewis, and Miller (2011) argue that cooperative education (the systematic alternation of school and work) creates ongoing partnerships between institutions of higher education and their corporate partners. These works offer a thorough look at the use of emerging technologies in university-community partnerships. However, a great deal of information, data, and analysis is still needed for further developments relating to these ideas. Interestingly, Whiteside (2003) investigates the skills that technical communication graduates and prospective employers think students need before entering business and industry. According to Whiteside, "Advisors, instructors, and professionals in business and industry should strongly consider the powerful nature of industry-academe partnerships" (p. 314). The question then becomes how to design different kinds of partnerships to maximize what participants can gain from such relationships.

Voss (2011), a leader in the design of partnerships between Lockheed Martin and the University of Central Florida (UCF), helped establish valuable connections that involve the university, Lockheed Martin, and the technical communication program. Voss' plenary presentation at the 2011 CPTSC annual conference focused on such partnerships. In that presentation, he referenced the success of recent alumni and their fruitful collaborations within the aerospace engineering industry. In addressing the topic, "Building Bridges for Industry and Academe: It's a Two-Way Street," Voss offered concrete suggestions for forging partnerships between industry and academe, including linking education and industry along with commercial enterprises. His research extends beyond the context of professors and industry partners to include, specifically, students at various levels.

The need to rethink and expand understandings of such partnerships has also become an increasing area of interest among academics, educators, and administrators. Rude (2011), for example, calls for additional research in academe-industry partnerships. In contrast to Voss' more industry-based perspective, her opinion is based on years of work in technical communication programs at Texas Tech University and Virginia Tech. Along these lines, Rude's recent research explores academe-industry partnerships as a significant theme in technical communication scholarship, a catalyst for much discussion in the field. In examining such partnerships, she identifies common themes of interest to both academe and industry—areas that could serve as a foundation for developing effective partnerships across the two kinds of organizations.

Like others, Bosley (1995) suggests that partnerships between technical communication programs and practicing professionals are crucial to developing a new educational environment for future technical communicators. The case study presented here supports and extends Bosley's primary argument by looking at how partnerships provide new educational opportunities. In this particular case, educational environments were constructed around the collaborative space of the partnership. In so doing, the relationship paired students with industry leaders and other students to work in service-oriented situations. These educational environments also serve as spaces where technical communication students can gain experience from mentors working in industry. They also provided students with the opportunity to serve as mentors themselves.

ENGAGING MULTIPLE POPULATIONS THROUGH EMERGING TECHNOLOGIES

The partnership documented in this chapter began as an extension of an existing relationship with the Madison County school district and their technology integration specialist. During the first session, a group of middle school students spent a day researching projects about local industry in general and on the BGAD in particular. For this process, students began their experience in the archives of EKU's Crabbe Library and then spent an additional day in the Noel Studio's Discovery Classroom. Within the context of such activities, EKU's Noel Studio provides an environment where students can use technology (e.g., touch-screen monitors and mobile devices such as Flip cameras and iPod Touches) to develop their technical communication skills such as structuring information, document design, and creating user-friendly texts.

The Integration of Technology

During the second session, middle school students were organized into small groups tasked with archiving the BGAD's history. They were asked to pay particular attention to that history as it relates to the operation's practices, activities, and relationships with the town of Richmond, for the BGAD and local historical society. The second session was also hands-on, providing students with access to and instruction on how to use technologies like Flip cameras, touch-screen monitors, and iPods. During this session, three Noel Studio staff members, in collaboration with three middle school teachers, instructed students to assemble their research in a way that would make sense to readers and viewers for years to come.

One challenge the students faced was making complex information (e.g. videos, maps, and supporting documentation) accessible. In response to this situation, the students took their audience to be the general population living near the BGAD. Students then articulated that the members of this audience would likely not have the technical background needed to effectively navigate documents. They, however, agreed that this audience would probably have a good general or historical knowledge of the BGAD. Students, for example, assumed their target audience would probably know information such as when the BGAD opened, the relationship between the town and the BGAD, and the health-related facts provided to the local community via BGAD reports. The student participants thus brought this local knowledge of the BGAD and its operations to the partnership as well.

For these projects, students focused specifically on forming partnerships with local industry, in this case the BGAD. To this end, small groups of middle school students worked with trained Noel Studio staff members on large touch-screen monitors. They also used technologies such as Google Earth to compare the proximity of nearby neighborhoods to weapons stores (noted by BGAD documents) and to the EKU campus. The students and Studio members also used historical geographic data (e.g., historical relationships between the city of Richmond and the BGAD) to determine the proximity of the BGAD to houses, business, and other public areas. During this process, students and Studio consultants used touch-screen monitors to locate and zoom in on reference points from different aerial and street-level views of certain areas.

The technologies noted here allow students to access different social and geographical reference points they could use to guide the communication design process. Students used these technologies to discuss the activities taking place at the BGAD as compared to those taking place in the surrounding areas. Additionally, students could engage in such processes while traversing the area via mapping software. Through such activities, students became instant audience members, and they also provided each other with supportive and at times objective comments throughout the process. Through this use of technology—mapping software on 42" touch-screen monitors—students were able to deepen their understanding of community-industry relationships by seeing how their projects drew from and touched upon various stakeholder groups. The geographical software also helped students visualize political and environmental issues (e.g., the potential environmental impacts of a chemical spill) in ways that they might not have done so otherwise.

Enhancing Connections to Community

The partnership outlined here involved students living in the community and whose families worked for local employers. In fostering and supporting this particular partnership, the Noel Studio served a far more substantial purpose than just providing an ideal space in which the partnership could occur or even serving as the catalyst for creating partnerships with the community. In reality, the Noel Studio served as the intellectual hub and inspiration for partnerships that

- Extend the reach of the campus community.
- Align complementary programs for a common goal.
- Engage community and industry through emerging technologies in a project of mutual benefit for all parties.

The content of the partnership sought to engage these organizations aligning for multiple purposes, including technical communication and community literacy.

One outcome of the project was that participants became more knowledgeable about area issues, biohazards, and local industry through the design of technical documents. The students and larger community, in turn, benefited from an objective source of information designed through the partnership—objectivity taught through research. For example, students discussed the difference between reliable and unreliable information as they planned their projects. In addition, the community became more literate[1] through the design of the communication resulting from this project, for it produced a source of archival research that the community could include as part of its repository of local information.

EXAMINING THIS PARTNERSHIP

Partnerships can serve valuable roles within the community, and they establish meaningful relationships among participants from technical communication programs, communities, and industry. Further, access to technology increases opportunities for communication-design experience within these partnerships and, in this case, participants' relationships with the community and industry around them. To examine such relationships in more detail, the author has conducted a case study of a partnership that assembled EKU, the BGAD, and a local middle school with the goal of teaching technical communication skills as students prepared communication documents for local industry that archive important information about the BGAD relevant for community members and historians.

The case study methodology employed in this chapter allows for rich description of the partnership with students from a local middle school collaborating with Noel Studio staff to produce a technical communication project documenting the history of the BGAD. This case study also allows for a more in-depth exploration and examination of both the approach to communicating somewhat technical subject matter and the collaborative relationship between the university and industry as reviewed in this case.

In performing this study, I analyze the case from multiple perspectives, including the community, industry, and university angles. Partnerships involve multiple stakeholders. With these additional voices, establishing programs that are productive for partners on both sides can be challenging. In this case, I use EKU, an institution with strong regional ties, to highlight the importance of engaging multiple stakeholders in partnership design.

Framework

As an approach to exploring partnerships, I advocate for a humanistic perspective, for it has been prevalent in technical communication theory and previous research in the field (see Rutter, 1991; Selber, 2004). I apply a humanistic

[1] For purposes of this chapter and the projects it describes, I am using the term "literate" to mean the ability to understand, analyze, and produce written, visual, oral, and multimodal information.

approach as a treatment, in this context, for understanding the design of academe-industry partnerships as one that is replicable across many contexts. This humanistic approach prioritizes learning and experienced-based outcomes in partnership efforts.

Following the work of Miller (1979) and Dombrowski (1994), I argue that the goals of partnership efforts are and should be human ends, the stakeholders involved in the partnership from the standpoint of communicators and clients. In the case described here (one that involves teaching students to deal with technical content in an interesting and interactive way), a humanistic approach suggests that even technical content such as the destruction of poisonous and even deadly gases and weapons should hold the people involved, including military personnel, students, and citizens, as central to the partnership. That is, I argue that the stakeholders involved should be central to decisions and outcomes.

In some cases, the partnership itself is as important, if not more important, than the content and outcome. That is, for technical communicators, partnerships provide the space and resources for students to see complex content and issues from multiple perspectives. In so doing, such partnerships allow students to take a role in designing documentation with the guidance of supportive and informed stakeholders. The result is real-world experience and perspective not available within the context of the classroom with no external stakeholder input.

To help readers better understand the various groups participating in this overall case, I have provided the following brief descriptions of the perspectives—or stakeholders—involved in the partnership studied for this case. Note that these descriptions are not intended to be exhaustive; rather, they simply add context for better understanding this partnership. I anticipate that, through the partnership-design process, a variety of collaborative models could be established.

Community

For this chapter, *community* is defined as the mix of people (e.g., citizens) and organizations (e.g., industry, K–12 schools, nonprofits, and organizations) that compose the area or region immediately around Richmond, Kentucky. Universities often engage in community partnerships with the intent to enhance or sustain an activity for community benefit. Gander and Rhyason (2011) argue that

> universities are under pressure to show significant returns on the public investment they represent by demonstrating that they have a tangible impact on real problems. They recognize that critical social problems can only be addressed through approaches that require academics to collaborate not only within disciplines but across them and to form partnerships with a range of external stakeholders. (p. 365)

Additionally, community involvement often adds local perspective to the partnership, thus complementing and extending industry and academe in ways that provide opportunities for individuals and stakeholders to make meaningful contributions, including when it comes to decision-making, for example (Adria & Mao, 2011).

Industry

Although the term *industry* can be broadly defined, within the context of this case, it encompasses the employment or business activity surrounding or near the EKU campus. Industry can include the public or private sector organizations and government or military entities. In the case described in this chapter, BGAD was engaged through the partnership with the university and serves as a major industry employer in Richmond.

University

The *university's* role in developing productive partnerships is perhaps the most complex to define. It is the academic institution within partnerships. Underwood and Parker (2011) argue that "the university needs to step forward as a more active institutional partner in these collaborative efforts," referencing their partnership model for University-Community Links (UC Links) (p. 311). Its role is often to add academic perspective, student researchers, or professional researchers to the partnership. Universities are often engaged for their intellectual capital, although this situation can create challenges when negotiating preferred outcomes for partnerships.

Within this overall context, a humanistic perspective preferences the needs and expectations of the individuals involved in the partnerships as outcomes are defined. For example, student learning was the primary goal for the partnership described here, with the secondary goal of producing user-friendly and informative documentation. This approach to partnership design might lead us to believe that people (i.e., the participants or beneficiaries) involved in the partnership should be given priority. Accordingly, outcomes should be established so that they contribute to the education, development, or quality of life of the area. In the case examined in this chapter, EKU and the Noel Studio served as the centralizing agents that brought together members of the community and area industry.

APPLICATION OF IDEAS

A substantial amount of scholarship explores service learning and civic engagement (Bowdon, Billig, & Holland, 2008; Bowdon & Carpenter, 2011), and others focus on the use of emerging technology (Bowdon & Carpenter, 2011). This chapter, however, highlights a partnership in a rural area of Kentucky where, historically, technical communication has not served a centralizing role. In rural areas such as this, partnerships around technical communication stand a better chance for success if they engage the people, communities, and industries of the region. In eastern Kentucky, for example, a great deal of industry is based on natural resources or the opportunity to develop industry in low-population areas. Industry, like the BGAD, is located in the area due to its vast landscape and rural setting.

The partnership discussed here provides us with an example of the potential for technical communication design to serve as a safe, inspiring, and supportive intellectual exercise for students to develop important skills. It does so by applying a humanistic perspective to an area of industry that remains somewhat mysterious to members of the community. That is, a humanistic perspective suggests that there is value in bringing entities together for the purpose of communication design, especially if the material is of interest to the community.

The Noel Studio is open to the university community and facilitates partnerships. However, the BGAD is an outside entity operating in a community that does not have input into its operations other than committee meetings. Information is provided to interested parties, but it is often produced internally and then disseminated for public consumption. While the BGAD has developed initiatives that involve community input such as advisory committees, much of its communication is designed from within. This partnership provided an opportunity for students to formulate new understandings of this local industry. Through this partnership, students experienced some of the challenges of collaboration when planning and designing technical documents.

IDEAS AND PRACTICES DRAWN FROM THIS CASE

As others design their own partnerships, I expect that these lessons learned from this case can be applicable across a variety of institutional and local contexts. Moreover, I expect that partnerships developed in rural areas like the one described here will look beyond the walls of their institutions and industries to involve other key stakeholders in the community. These partners might include local schools, community literacy programs, or public services that extend the reach of the university. This humanistic approach to partnership design situates stakeholders as central to these arrangements, which provides opportunities for education that links academic and professional perspectives on technical communication. This partnership design suggests that

- Academe-industry partnerships can benefit both sides.
- Partnerships can provide the opportunity to teach technical communication to students at a variety of levels.
- Partnerships that link industry and academe can include stakeholders from the community.

Such a framework can provide individuals in technical communication with a mechanism of how to reconsider and reconceptualize how organizations representing the academy and industry might engage in more effective collaborations.

In the end, it's a matter of fostering effective engagement across participating stakeholder groups.

Academe-industry partnerships must engage the people and local industry in their communities in order to be successful. This engagement means creating relationships, opportunities, and collaborations that consider both K–12 and industry, seeing the university as a catalyst for partnerships. With this perspective in mind, I suggest that a humanistic rationale for technical communication can also inform the design of academe-industry partnerships. A humanistic rationale gives perspective to the students involved, creating an opportunity for further investigation into similar partnerships that exist either in metropolitan areas—where partnerships thrive at major universities—or more rural areas like Richmond, Kentucky.

BEST PRACTICES FOR USING A HUMANISTIC APPROACH TO CREATE EFFECTIVE COMMUNITY PARTNERSHIPS

In regions such as eastern Kentucky, technical communicators must carefully examine the intersections of community involvement within the area to discover partnership opportunities. I argue, though, that a humanistic perspective for partnerships provides students with leveling opportunities, in this case, to document an important aspect of the area's history. A humanistic perspective, as Miller (1979) and Dombrowski (1994) have argued, brings technical communication back to the purpose of communication itself: to make technical information accessible.

The BGAD, however, is a complex client with serious subject matter. The website for the U.S. Army Chemical Materials Agency (2011a, 2011b) conveys a technical look at the subject of chemical destruction. The Web content suggests a serious tone, with complex and inaccessible subject matter. The content discussed during the collaboration explored the complex and serious nature of the subject matter, at times using the available technology to access maps and data. Conversations focused on the risks associated with having such an installation so close to residential neighborhoods and a university. In the case examined here, many of the students live in the same area (within a mile or two) of the BGAD. By the very proximity of the industry, the partnership takes into account the potential hazards to the community, natural resources, and the people living in the area.

The design of this partnership attempted to engage multiple community stakeholders—the university, the BGAD, and local K–12 students—in a collaborative technical communication experience. The subject matter was to draw from the history of the BGAD within the community. Students were to document this history using technology and their growing skills as communicators. To say the least, students were excited about leaving the classroom for this project and engaging spaces of the local university. The partnership accomplished all of these goals, yet the subject matter of the partnership itself, as is the case at many other universities, is quite technical and serious in nature. The humanistic perspective proposed here considers the technical and serious nature of the activities taking place in the BGAD but looks deeper to consider

- The activity of the partnership
- The value of technical communication education and experiences for the community
- The potential for technical communication partnerships to add value and knowledge to all sides of the community, from K–12 to higher education and industry

Understanding and addressing such factors can help a range of organizations better engage in partnerships across the academy and industry.

The humanistic lens applied here takes into account the challenges of developing partnerships between the university and industry in rural areas. While there are a number of well-documented challenges to establishing partnerships, the potential they offer for students to gain useful technical communication perspectives cannot be overlooked. Furthermore, in the relationship described here, students on the university side also serve important leadership roles, an opportunity they would not have had without the partnership.

Hayhoe (1998) asks, "What's in it for all of us?" I would suggest that the answer to this question is rather complex and depends on a number of cultural, social, and regional considerations. One way to answer this question is through a humanistic perspective for the design of academia-industry partnerships, showing at its heart that partnerships serve the area and community in addition to the university. A humanistic perspective also encourages participation from multiple stakeholders—the university, industry, nonprofits, and K–12. Thus, partnership initiatives are intensified and stratified, benefiting multiple populations.

Exploring challenges in partnership formation within the region served by EKU, a humanistic perspective works especially well in rural areas where it is challenging to align partnerships. Building on the many compelling and productive partnerships established through technical communication programs, this chapter attempts to refine perspective on the design of future initiatives.

CONCLUSION

The humanistic perspective applied here has helped the Noel Studio establish partnerships that engage community populations, including K–12 and industry. In fall 2011, the same semester that this partnership was conducted, two new studios at local high schools emerged as evidence suggesting that a space and vision for partnerships can have tremendous, positive, and long-lasting results.

Employing this perspective has also yielded opportunities to partner with local industry and to develop cohesive efforts to promote technical communication practices. A humanistic perspective can help academe-industry partners develop goals and relationships that transcend individual interests and fully engage all sides.

REFERENCES

Adria, M., & Mao, Y. (2011). Encouraging public involvement in public policymaking through university-government collaboration. In M. A. Bowdon & R. G. Carpenter (Eds.), *Higher education, emerging technologies, and community partnerships: Concepts, models, and practices* (pp. 374–380). Hershey, PA: IGI Global.

Bosley, D. (1995). Collaborative partnerships: Academia and industry working together. *Technical Communication, 42*(4), 611–619.

Bowdon, M. (2011). Preface. In M. A. Bowdon & R. G. Carpenter (Eds.), *Higher education, emerging technologies, and community partnerships: Concepts, models, and practices* (pp. xxviii–xxxvi). Hershey, PA: IGI Global.

Bowdon, M. A., Billig, S. H., & Holland, B. A. (Eds.). (2008). *Scholarship for sustaining service-learning and civic engagement*. Charlotte, NC: Information Age.

Bowdon, M. A., & Carpenter, R. G. (Eds.). (2011). *Higher education, emerging technologies, and community partnerships: Concepts, models, and practices*. Hershey, PA: IGI Global.

Bowdon, M., & Scott, B. (2003). *Service-learning in technical and professional communication*. New York, NY: Allyn & Bacon.

Cates, C., Cedercreutz, K., Harfmann, A. C., Lewis, M. W., & Miller, R. (2011). Using digital technology to enhance a century old partnership between university and cooperative education employers. In M. A. Bowdon & R. G. Carpenter (Eds.), *Higher education, emerging technologies, and community partnerships: Concepts, models, and practices* (pp. 350–358). Hershey, PA: IGI Global.

Culliton, B. J. (1983). Academe and industry debate partnership. *Science, 219,* 150–151.

DeBerg, C. (2011). SAGE: An international partnership linking high school students, universities and the private sector through social enterprise. In M. A. Bowdon & R. G. Carpenter (Eds.), *Higher education, emerging technologies, and community partnerships: Concepts, models, and practices* (pp. 15–28). Hershey, PA: IGI Global.

Dombrowski, P. M. (1994). Humanism and technical communication. In P. M. Dombrowski (Ed.), *Humanistic aspects of technical communication* (pp. 1–13). Amityville, NY: Baywood.

Engineering Advisory Committee Subcommittee on Industry-University Partnerships. (2008). *Encouraging industry university partnerships.* Retrieved July 30, 2012, from http://www.nsf.gov/oirm/bocomm/meetings/nov_2009/EAC_UIP_report.v4.pdf

Gander, L., & Rhyason, D. (2011). Community-university engagement in an electronically-defined era. In M. A. Bowdon & R. G. Carpenter (Eds.), *Higher education, emerging technologies, and community partnerships: Concepts, models, and practices* (pp. 365–373). Hershey, PA: IGI Global.

Hayhoe, G. (1998). The academe-industry partnership: What's in it for all of us? *Technical Communication, 45*(1), 19–20.

Henry, J. (2011). Hybridizing F2F and virtual collaboration between a government agency and service-learning technical writing students. In M. A. Bowdon & R. G. Carpenter (Eds.), *Higher education, emerging technologies, and community partnerships: Concepts, models, and practices* (pp. 58–67). Hershey, PA: IGI Global.

McEachern, R. W. (2001). Problems in service learning and technical/professional writing: Incorporating the perspective of nonprofit management. *Technical Communication Quarterly, 10*(2), 211–224.

Miller, C. (1979). A humanistic rationale for technical writing. *College English, 40,* 610–617.

Rude, C. (2011, October). *Building bridges for industry and academe: It's a two-way street.* Plenary speech presented at the Council on Programs in Technical and Scientific Communication conference, Harrisonburg, VA. Retrieved August 15, 2012, from http://www.cptsc.org/annual/cptsc-2011/plenary-speakers.pdf

Rutter, R. (1991). History, rhetoric, and humanism: Toward a more comprehensive definition of technical communication. *Journal of Technical Writing and Communication, 21*(2), 133–153.

Santoro, M. D., & Betts, S. C. (2002). Making industry-university partnerships work. *Research Technology Management, 45*(3), 42–46.

Selber, S. (2004). *Multiliteracies for a digital age.* Carbondale, IL: Southern Illinois University Press.

Society for Technical Communication (STC). (2011). *Strategic goals.* Retrieved August 15, 2012.

Starke-Meyerring, D., Duin, A. H., & Palvetzian, T. (2007). Global partnerships: Positioning technical communication programs in the context of globalization. *Technical Communication Quarterly, 16*(2), 139–174.

Tovey, J. (2001). Building connections between university and industry: Implementing an internship program at a regional university. *Technical Communication Quarterly, 10*(2), 225–239.

Underwood, C., & Parker, L. (2011). The tools at hand: Agency, industry, and technological innovation in a distributed learning community. In M. A. Bowdon & R. G. Carpenter (Eds.), *Higher education, emerging technologies, and community partnerships: Concepts, models, and practices* (pp. 300–313). Hershey, PA: IGI Global.

U.S. Army Chemical Materials Agency. (2011a). A job well done. *CMA.* Retrieved August 15, 2012, from http://www.cma.army.mil/home.aspx

U.S. Army Chemical Materials Agency. (2011b). Richmond, KY. *CMA.* Retrieved August 15, 2012, from http://www.cma.army.mil/bluegrass.aspx

Voss, D. (2011, October). *Building bridges for industry and academe: It's a two-way street*, Plenary speech presented at the Council on Programs in Technical and Scientific Communication conference, Harrisonburg, VA. Retrieved August 15, 2012, from http://www.cptsc.org/annual/cptsc-2011/plenary-speakers.pdf

Whiteside, A. (2003). The skills that technical communicators need: An investigation of technical communication graduates, managers, and curricula. *Journal of Technical Writing and Communication, 33*(4), 303–318.

CHAPTER 8

Accommodationist and Constructivist Approaches to Academic-Industry Partnerships in a Usability and User-Experience Facility

Tharon Howard

ABSTRACT

This chapter provides a retrospective of the academic-industry partnerships that emerged from a usability testing facility begun in 1993. The chapter details how the facility began and describes how, over two decades, the facility took a 180-degree turn from what the author calls an "accommodationist approach" to a "constructivist approach." The author describes how, in response to the changing needs of his industry partners and the evolution of the usability testing field, the research focus of the Usability Testing Facility changed from helping industry partners accommodate technologies to the needs of customers (the accommodationist approach) to constructing social roles users can play to take advantage of the technologies (the constructivist approach). The chapter also provides insights into strategic choices the facility's director made to develop sustainable, long-term partnerships with industry clients.

INTRODUCTION

In their 2011 reflective essay, Ginny Redish and Carol Barnum trace the history of usability testing and user experience design from the late 1980s through the

2010s and describe some major organizational developments in our field. Entitled "Overlap, Influence, Intertwining: The Interplay of UX and Technical Communication," Ginny's portion of the article maps out the beginnings of the Usability Professional Association (UPA) in 1993 and the start of an influential online community called UTEST, which I also began in 1993.

As Ginny observes, in those early days when usability was just beginning to affect the field of technical communication, few works aside from Neilson and Mack's *Usability Inspection Methods* (1994) or Dumas and Redish's *Practical Guide to Usability Testing* (1993) existed to guide the practicing professional. As a result, our field's research methods were fairly limited. Think-aloud protocols and heuristic evaluations were the norm, and the UPA, which was started by Janice James and the UTEST community, were primarily populated by researchers seeking more robust and rigorous methods of conducting usability research.

Rolf Molich's (1998) famous CUE studies give some indication of how methodologically naïve we were in these early days of usability testing. In his CUE studies, Rolf compared the research findings and recommendation reports of usability professionals confronted with the same research questions and the same data. Rolf found, despite the fact that usability professionals were starting with the same problems and the same information, that the findings and recommendations from one research team to another were often startlingly different. This lack of replication by other usability researchers doing essentially the same study was disturbing evidence of the field's immature and sometimes eclectic approach to the study of usability. In response, professional organizations like the UPA and online communities like UTEST undertook efforts to legitimize usability as a scientific mode of inquiry that sought to push usability testing beyond merely subjective responses and anecdotal feedback produced by focus groups and less rigorous forms of market research.

The Clemson Usability Testing Facility (UTF) came into being against this backdrop of methodological immaturity and the struggle for professional legitimacy. Officially started in 1993, the Clemson UTF was part of an academic-industry collaboration between NCR Corporation and the Master's of Arts program in Professional Communication (MAPC) at Clemson. As a land-grant university charged with providing outreach services, it was appropriate that the MAPC faculty were asked by industry to provide usability testing services for their documentation sets and information products. Initially, the MAPC faculty began this work by performing usability evaluations as client-based projects in our undergraduate technical writing courses. However, the "hand-to-mouth" nature of those projects was unsustainable, and so we began the development of a standing resource—a laboratory—where we could respond more nimbly to the requests of our industry partners. This initial laboratory became the UTF, which I have directed since its inception in 1993.

This chapter provides a high-level overview of how the UTF has evolved over the past 20 years. More specifically, I will examine how the academic-industry partnerships developed in the UTF have been molded and shaped by the evolution of the field of usability testing and user-experience design itself. I will also show how the field of usability testing and, by extension, the UTF have undergone three phases of development I call

1. The accommodationist phase
2. Usability puberty
3. The constructivist phase

For each phase, I will discuss how the UTF's relationship with industry partners underwent changes in the research questions we were asked to investigate, the types of data we collected, and the nature of the recommendations we reported to our industry partners.

THE ACCOMMODATIONIST APPROACH

Because it was so new and its methods so limited, it is not surprising that the usability testing of the early 1990s tended to focus on capturing the "low-hanging fruit" in product development and assessment. In those early days, we generally adopted an "accommodationist" approach to our work. We mainly sought to produce research that showed product designers how they could "accommodate" the technologies, products, and services they were developing to their users. At the time, usability testing professionals primarily did this by showing designers the "mistakes" that they had made in their functional beta units. At this same time, the MAPC program recruited new industry partners to work with our usability testing facility by persuading them that users shouldn't have to adapt to the technological products that industries were producing. In so doing, the MAPC program encouraged our corporate partners to become "user advocates" who accommodated technologies to meet users' goals and needs.

Generally speaking, this accommodationist approach was (and for that matter still is) extremely successful. But in 1993, it resonated particularly well with the MAPC's industry partners in the high-tech sector since the computer industry was emerging from the days when manufacturers had been distinguishing themselves from the competition with what were affectionately known as "speeds 'n feeds." Today's PC consumers pay relatively little attention to the clock speeds of their CPUs. Though it's not irrelevant, speed has become a relatively minor concern these days as there are a wide variety of competing systems all running at roughly similar speeds. However, in the late 1980s, speed was king. If a manufacturer could produce a chip that eked out a few more MHz than the competition, there was a good chance that systems with those chips would become the market leaders, regardless of most other purchasing considerations.

But as more chip manufacturers entered the market in the 1990s, functionality concerns like speed and technical specifications played a less dominant role in the decision triggers that drove purchases. Consequently, in the early 1990s, manufacturers were already seeking alternative ways to differentiate themselves from the competition. They were seeking ways to make ease of use, compatibility, and simplicity play a more significant role in product design at the same time usability practitioners were developing more rigorous and reliable research methods and establishing their identity as a profession.

In hindsight, it is easy to see that the convergence of usability testing's professionalization and the end of the speeds 'n feeds era in technology development meant most of the early academic-industry partnerships in my lab were with companies like NCR, IBM, and AT&T GIS. Yet, the UTF's main entry into those partnerships didn't start by selling our usability testing services to hardware design teams at those companies. Instead, our contacts developed by trying to help companies develop more "user-friendly" manuals and documentation sets. At the time, the project managers at these companies were looking for ways to help technical writers improve their companies' customer support and information products so the companies could gain a competitive advantage in the market. The technical writers, in turn, were looking for help learning how to move their documentation from paper manuals into online help, and we at the lab were looking for opportunities for our students to work with corporate clients. The timing couldn't have been more perfect.

In response to this situation, my MAPC colleagues and I began to forge relationships with industry by having students in our technical writing courses work with information product developers at NCR to produce manuals and documentation. As part of the course, we required students to conduct usability testing on the manuals, and products were being used and developed in the course. Not surprisingly, the students' reports revealed important design limitations even good documentation couldn't address. As a result, the NCR managers were predisposed to respond positively when we approached them with a proposal to develop a stand-alone usability testing facility on the Clemson campus. Because the facility was offsite, NCR's management could argue that our findings were more objective than in-house studies could provide.

One example of how this relationship worked and how it was dominated by the accommodationist approach can be seen in the first study we conducted for NCR. At the time, NCR's primary market for computers consisted of enterprise clients rather than the "SOHO" (small office/home office) market or the general public. NCR sold most of its workstations and servers to large corporations, state and federal government agencies, or engineering firms, so its enterprise computers weren't being purchased by the end user. Instead, these computers were acquired by agents in a special purchasing office, and these offices generally have criteria for purchasing decisions (e.g., concerns about the upgradability and maintenance costs of the systems). As a result, if NCR could demonstrate to enterprise buyers that the NCR systems would cost less to maintain and were easier and less costly to upgrade over a typical 3-year upgrade cycle, then NCR could gain competitive advantage in the enterprise market.

NCR sought to make their computers easier to maintain and upgrade by designing an entirely new architecture for their motherboards, an architecture called "microchannel." To put this in context, it's important to remember just how difficult it was to change and install a new video card, modem, or sound board in a PC's motherboard 20 years ago. In those days, Windows 3.1 still ran on top of DOS, and new cards and devices weren't "automagically" recognized by the system when plugged into the motherboard. Installing a new card in most PCs of the day required that users understood how to configure the operating system to avoid IRQ (interrupt request) conflicts between specific devices, how to resolve input/output base address conflicts, how to install the device's software drivers, how to configure Windows so it could find the dynamic link library files (aka .dll's) and many other issues. The NCR's microchannel architecture was an early effort to resolve these problems, and it worked the way USB devices function today. Plugging in a microchannel-compliant card in an NCR system of the day eliminated many concerns because the motherboard recognized the device once it was connected.

For NCR's enterprise customers, microchannel meant they could reduce technical support costs as less need existed for a large IT support staff to maintain and upgrade systems. Moreover, by making it easier for the staff to replace or repair cards, they could work more efficiently when engaging in such processes. However, NCR wanted to test the system to ensure the designers and engineers hadn't overlooked potential problems—something our previous studies in the technical writing classes demonstrated usability testing could do. Consequently, one of our first usability studies for NCR involved IT professionals using an NCR 3360's manual to try to install microchannel cards into an NCR 3360 personal workstation.

And as one might expect, our study revealed there were a number of issues designers hadn't anticipated. For example, we discovered that an entire section of the user manual was inaccurate and would mean customers who attempted to use it would eventually have to call NCR's customer support for help. Having an external research group conduct the usability study revealed these kinds of problems. Moreover, the project manager for the team estimated that this single usability-related finding saved them $58,000 in printing costs because graduate students working in the UTF caught the problem in time to stop a major printing run of the defective manuals. Additionally, the UTF's report was able to demonstrate that this usability study saved the company substantial costs by showing the information product developers how to revise the documentation. As a result, far fewer consumers would need to telephone customer support.

What is important about this initial study is it reveals some of the classic problems of the accommodationist phase. What is noticeable about the clients in the early days of the UTF is researchers were being asked to evaluate the usability of a product to demonstrate the product was accommodating the user's needs. In the NCR 3360 study, we were *validating* ideas that originated elsewhere—Marketing, Engineering, and Management. We were not testing in order to develop new products, and we were not studying the work environment in order to identify tasks and goals users had in that environment. We were instead engaged in a usability "bug-fest" wherein we found flaws in other colleagues' designs.

During this phase, we had to learn that it was critical to the long-term success of our partnership that we had to show how designs had failed to accommodate users and invest equal amounts of effort to show how products successfully met users' needs. We learned that the long-term success of our partnerships were based on helping engineers, designers, and writers feel we weren't being brought in by management to evaluate *them*. For this reason, we worked hard to make sure we worked *directly* with the people who developed the products to improve the *product* rather than working to find mistakes in developers' work.

To counter blowback from the accommodationist approach to usability, I developed partnership-building techniques that worked against the impression that the UTF was identifying developers' mistakes. For example, I made sure our contracts with industry partners explicitly stated our recommendation reports would be made directly to members of the product development team. My contracts stipulated that only one person on the product development team would serve as the "liaison" between the corporate partner and me. The liaison was responsible for all communication from the company, so if, for example, the vice president of marketing wanted to change the types of customers for whom the product was being designed, that information went through the liaison. The VP couldn't come directly to me and tell me I needed to recruit a different demographic for participants in the usability studies. In this way, we made sure that the product development team wasn't left out of the loop.

These strategies also provided a means for protecting graduate students who worked on the UTF team from potential abuse. The contracts not only stipulated information flow through the liaison on the company side, they also stipulated that information would flow through me on the UTF side. Thus, if an employee in the company asked one of the graduate students on the team to make a change in our research design, the student could legitimately say she didn't have the authority to make such a change.

Just as the contract ensured information from the company came directly to me, the contracts further stipulated that the UTF's reports would go only to the liaison to be distributed by the liaison. This action prevented management or employees from other units from coming directly to me and demanding to see copies of the UTF's reports. This protected the UTF team from the political games that sometimes occur in corporate environments. Because the liaison was the only person who could distribute the reports, the UTF team didn't have to provide a copy of our research to any employee who requested it, including developers on other projects who might be competing for resources with the team that our liaison supported.

More than any other strategy we used, this single stipulation in our contracts was perhaps the most important because it enabled the product development team to exercise considerable control over how the UTF's reports were used. This control over the report's distribution was a critical aspect of getting the developers to trust me and the UTF team. Naturally, there were times when I regretted that the UTF's reports weren't going directly to upper management (particularly when the team chose not to follow some of my "best" recommendations). Nonetheless, having the trust and support of the people working directly on the products being developed and tested was ultimately more important to the success of our testing and to whether our recommendations were implemented in the final product design.

One last strategy that mitigated against the blowback from the accommodationist approach involved making sure the scope of work in our contracts explicitly stated that we would *recommend only changes* to a product design; we would not actually be responsible for making or implementing those changes. This stipulation was important, for it countered the impression among developers that management had "hired a bunch of high-powered PhDs and academics to swoop in and take over our jobs." Because of this stipulation, we could legitimately tell developers that the UTF team was there only to provide recommendations aimed at helping *the developers* to improve their products. We couldn't take over their project or wrest ownership of the product away from them, for it was entirely up to the developers to decide whether or not to implement our recommendations. This fact, combined with the strategy of giving the development team control over how our recommendations were distributed, helped in building sustainable usability testing partnerships.

Despite our best efforts to secure projects that would bring us into the development process early enough to help innovate or to discover new product ideas, our clients essentially wanted us to help them prove (usually to management) that the information products they had designed were or were not usable for a particular audience demographic. Indeed, our clients were often project managers who were frustrated by their inability to convince upper management that they needed more resources to make documentation, instructions, training, and packaging more effective. These clients would request usability testing knowing full well that the results we provided would demonstrate the need to invest resources in a new or a significantly revised information delivery systems. In a surprising number of cases, our clients came to us because they could not be "prophets in their own land." They needed the voice of an independent organization to get upper management to realize the documentation set and information delivery systems being developed were affecting customer support and satisfaction.

In one case, a corporate vice president decided the design team needed to consider using a "multimedia keyboard" with built-in audio components, thus eliminating the need for telemarketers to wear headsets. The idea was that technology would simplify the setup of the system, reduce the telemarketer's required workspace, and reduce production costs by eliminating the need for separate ports for speakers and microphones. The design team knew a multimedia keyboard would be completely unacceptable. The designers already knew that few telemarketers work in private, enclosed office spaces, so the noise that external speakers would make would be extremely disruptive. Nonetheless, rather than confronting the vice president with what the designers already knew, the design team's project manager called in the UTF to conduct a task-performance study with competing multimedia keyboards to get users' feedback on the competing multimedia keyboard designs.

No one on the design team told me about the corporate involvement in the situation and the real reason for which we were testing the keyboards. My team and I also were not told that vendors who wished to secure contracts to provide such keyboards would be attending the meeting at which we presented our results. As a result, my team and I inadvertently antagonized the vendors who probably came to see "usability testing" as a form of summative evaluation. Because the usability testers were "laying down cover fire for a decision that had already been made," the situation hid the potential role in which *formative* usability testing could play in product development. Consequently, the "accommodationist" approach was all we needed as the research questions our clients wanted us to answer did not require other modes of inquiry.

USABILITY ENTERS PUBERTY

The accommodationist approach has never gone away; it's still an incredibly useful approach that *should be employed* during the development of new products and services. Over time, however, high-tech industries have become increasingly sophisticated. My clients in the usability lab are rarely seeking the services they were back in the early 1990s. They still ask for think-aloud protocol analyses to validate whether or not a product or service accommodates the goals and needs of a particular user demographic. Yet, at the same time, our clients now are equally interested in on-site needs-assessment studies in which my students and I examine users' work environments to identify opportunities for new products. Our industry partners now seek help with questions involving diary studies. These involve users sending us images and feedback chronicling 6 weeks of their experiences with a product, for such information helps us better understand the environmental factors influencing how and why they use a product.

Today, we have entered what I like to call the "constructivist" approach to usability and user experience design. Indeed, one could argue the term "user experience design" marks an effort by usability practitioners to break out of the "accommodationist" paradigm. It is a means to begin thinking of ways of constructing user roles and interpretive frameworks—ways that positioned users to effectively take advantage of interfaces and technologies. Where the "accommodationist" view helped designers affect technologies to accommodate users, the "constructivist" view uses social interfaces to adapt users to the technology.

One example of this constructivist move was the introduction of Microsoft's BOB interface in 1995. BOB was important because it tried to create a socially informed user position that gave the individual user power over the file manager, calendar, checkbook, and system navigation processes. It did so without forcing the user to learn arcane and uninteresting details of the operating system. BOB tried to do this by re-creating the living room, office, bedrooms, and others spaces in a traditional, suburban U.S. household that belonged to a character named Bob. Running as a shell on top of Windows 3.1, BOB eliminated the need for users to learn to navigate Program Manager. Instead, users would click on items such as a calendar on the wall to launch the scheduling application. This approach socialized users into a digital world where the social roles being played provided clues about where to look if one wanted to write a letter, balance a checkbook, update your contact list, or perform other routine household tasks.

Microsoft pulled BOB from the market within a year, and in 2010, *Time* listed it as one of the 50 worst inventions of all time (Fletcher, 2010). BOB was certainly a gigantic commercial failure, which the popular media loved to critique. Indeed, about the only time BOB gets mentioned today is when someone notes that Comic Sans font became a default true type font for Windows systems because it was originally designed for BOB. However, BOB deserves more credit than it has received, for it was a bold attempt to engineer human behaviors rather than accommodate them. BOB exemplified usability reaching puberty, and it was an awkward stage in our field's development. But as an experiment in constructing user behaviors rather than accommodating them, the interface deserves more credit than it has received.

While BOB represents one of the most public efforts to grow beyond the accommodationist approach, Microsoft wasn't the only company attempting such a move. In fact, such approaches led to a new range and kind of partnerships between our lab and a number of companies. In the late 1990s, the UTF partnered with IBM's Ease of Use and Architecture group to conduct a range of usability tests involving 3-dimensional interface designs that could potentially provide users with an interpretive framework granting them what I called "predictive power" over the interface. Essentially, predictive power meant a user could predict, with a high degree of accuracy and comfort, that interacting with an object in the interface would produce a specific result; for example, if as a user you click button A, then you can reliably predict that you will produce result B.

Users' memories obviously play a huge role in this process, and rhetoricians who have studied the classical art of memory know that *mnemotechnics* are intimately tied to 3-dimensional spaces. In the *Rhetorica ad Herennium*, for example, classical speakers learned to memorize extremely long speeches through the use of locations (Latin *loci*) in an architectural structure or some other 3-D space. These spaces were then memorized by practitioners as the framework for the speech. Speakers then mentally walked through a 3-D space, and as they entered each room, or *loci* in the space, they would "see" an object that would visually remind them of the content they were supposed to cover in the speech (see, for example, Cicero, 1955, pp. 205–225). So when someone says, "In the first place," "In the second place," and so on in modern speeches, it's a throwback to the ancient art of memory when speakers used 3-D tours in their minds to locate the order and content for their speeches.

Our idea for IBM was that you could hear the same phenomena operating when users were trying to "locate" information on their computers. Regardless of the operating system, when users would seek a file, they would ask questions like "Okay, so now *where* did I *put* that file?" In other words, a spatial logic seemed to be at work when many of our study participants performed routine file management and retrieval tasks, and we wanted to try to tap into it.

Worth noting is at this same period (1996), id Software's video game *Quake* became a huge hit on the Internet, and customer interest in building interfaces with 3-D real-time rendering developed. Consequently, we started investigating the desirability and feasibility of building a 3-D interface that gave users enhanced predictive power over the operating system and the routine business tasks typically found in SOHO environments. We also began designing research methods and studies that let us address new architectural problems we encountered. For example, what way-finding techniques were most effective for users in 3-D landscapes? Should verisimilitude drive the design of the environment, or should walls be transparent so users could walk through them? Should we turn off gravity so users could fly? And a particularly vexing question was how to map 3-D actions onto a 2-D input device like a mouse. For example, when users rolled the mouse forward, should the users move forward in the space or should they fly up? And how should they change states so the mouse could control both movement in the space and be used to interact with objects in the space on the other?

Such research questions stretched the limits of the traditional card-sorting and think-aloud research methodologies we had been using in the early 1990s. The rise of the World Wide Web and designing interfaces for that new environment pressed usability professionals to create new methods for studying the usability of these new environments (just as mobile computing is stretching us today). Yet the approach we were taking was still accommodationist at its core, even with the previously noted 3-D studies. True, we were applying usability to innovate at last, but we were still trying to find ways of accommodating the technologies to the users rather than trying to "construct" the user. Taking the 180-degree turn from accommodating technologies to users to constructing users who can accommodate the technologies wasn't something that happened over night. Still, I believe one major factor that brought about this change was the evolution of the research methods and the questions that usability researchers are asking.

Ultimately, a number of issues eventually shook usability professionals out of that evaluative mode and began forcing us to let go of the accommodationist-only paradigm. One issue is that we weren't innovating new products; we were only fixing old ones. Second, usability professionals were still often forced to work with functional prototypes, or the products were pretty far along in development before usability testing with think-aloud protocols could be implemented. For our industry partners, having to wait for high-fidelity prototypes seemed like we were slowing down the product development cycles because we were constantly recommending revisions to the prototypes. A third problem was pushback from the product developers who hated the idea that their work was being evaluated all the time. A final problem was a desire to find innovative, less costly ways to develop new products.

If we were to really understand users' work environments in order to address these problems, usability professionals required better needs assessment, task-analysis, diary study, and ethnographic/contextual analysis tools than were commonly used in the early 1990s. Finally, as usability testing and think-aloud protocol analyses became more mainstream modes of inquiry and were implemented by more technical writing teachers, the need for specialists with expertise in that particular skillset was reduced. As a result, usability professionals began seeking additional, alternative, and more sophisticated means of doing research to distinguish themselves from "weekend warriors," who thought they could do usability testing because they had taken a class in it. In terms of developing and maintaining our academic-industry partnerships in the UTF, this mainstreaming of the accommodationist modes of inquiry meant we too had to differentiate our services so that we were also offering skills our industry partners couldn't find among current employees. If the UTF wanted to continue enjoying successful partnerships, it became clear by the end of the 2001 "dot com" crash that the UTF needed to innovate its way out of the accommodationist paradigm.

CONSTRUCTIVIST SHIFT

The problem with describing the history of the constructivist shift is that the full-on user-experience-design perspective doesn't have a single moment in time when it appeared. Instead, the best one can say is the awareness that "usability" and "user-centered design" had become imbricated with "user-experience design." For many people in the usability community, that awareness happened 2 or 3 years after the dot com crash of 2001. In the UTF, we realized we weren't doing think-aloud protocols for industry partners nearly as often as in the past and recognized that other research methods in our tool box had become more important to our relationships with industry partners than traditional usability testing methods.

The UTF team soon realized we were examining different kinds of user concerns and playing different roles with respect to our clients. Of course, the UTF team still gave industry partners lists of recommendations for changes they needed to make. These changes were based on observations of users actually working with prototypes and products; however, they were only one of the deliverables we were providing our clients. More frequently, we did needs assessments wherein we described the functional, organizational, environmental, and even aesthetic needs of our clients' customers. We were also providing personas of our clients' customers to help those clients make decisions about the features that their products should or shouldn't offer. Most importantly, we were studying the customers of our industry partners to help clients make informed decisions about how design interfaces that created user experiences that *persuaded* users to engage in specific types of behaviors.

A review of some of the books recently published in the field suggests our experiences in the UTF at Clemson weren't unique. Stephen Anderson's *Seductive Interaction Design* (2012), Ian Bogost's *Persuasive Games* (2010), Robert Cialdini's *Influence: The Psychology of Persuasion* (2006), B. J. Fogg's *Persuasive Technology: Using Computers to Change What We Think and Do* (2002), Jeff Johnson's *Designing with the Mind in Mind* (2010), Don Norman's *Emotional Design: Why We Love (or Hate) Everyday Things* (2005), Daniel Pink's *Drive: The Surprising Truth about What Motivates Us* (2011), Trevor van Gorp and Edie Adams's *Design for Emotion* (2012), and Susan Weinschenk's *100 Things Every Designer Needs to Know about People* (2012) and *Neuro Web Design: What Makes Them Click?* (2009) represent a few of the books concerned with creating user experiences that alter the behaviors of users.

This shift in perspective has forced us to reassess our relationships with our clients and reconsider what services we provide. As Keith Grant-Davie (1992) observes,

> [When] researchers do not find exactly what they expected . . . researchers must negotiate with the data, searching their memories for alternative schemas (patterns of relationships into which people organize their knowledge) that might account for the data, revising the schemas they had brought to the analysis, or forming new schemas to account for the evidence. (p. 273)

Thus, even though one would expect a lab called the "Usability Testing Facility" to be conducting traditional usability studies like think-aloud protocols, the UTF was doing very few of them. Rather, once we realized our clients' goals were changing and we needed to adapt our research practices to sustain our relationships, we began seeking "alternative schema" to help explain the changes we observed. To that end, we found tremendous explanatory power in Louis Althusser's (1971) concept of "interpellation." I realized our research at the UTF during this decade attempted to find social, aesthetic, and cultural characteristics of the users of a particular product. And we were pursuing data about these characteristics so our industry partners could use that information to design interface cues. These cues would then interpellate users into a social role with an interpretive framework that gave them predictive power over the interface.

Althusser coined the term "interpellation" in 1971 to describe how people are "hailed" or assimilated into societal groups. As I explained in *Design to Thrive* (Howard, 2010), social groups have behavioral protocols that dictate acceptable interactions, so being interpellated into a group means you understand the conventions of the group and (consciously or not) agree to follow them. When humans behave in a way consistent with the rules of the community, their behavior results in a positive and reinforcing feeling of belonging to the group.

A simple example I often use to explain this phenomenon is the traditional business handshake. If I hold my right hand out to you, I have created a "gap" (Iser, 1978) in the greeting experience. I expect you to fill this gap by extending your own hand and grasping mine in the expected way (i.e., the way society expects you to respond). If you recognize the gap and if you respond by extending your hand, then you will have "completed the form" (Burke, 1968) of a Western-style greeting. By filling in the gap in the social protocol, you are interpellated into the group of people who are accustomed to saying hello in that manner. In short, interpellation is about recognizing gaps and giving in to the desire to fill them in socially acceptable ways. The result is much the same as the rhetorical effect of an enthymeme wherein a listener fills in the "gaps" or items not stated by the speaker. If, for example, I say to you that "All humans are mortal, and Socrates is human," then I don't have to fill the gap (i.e., the missing information/conclusion) for you—you know the form already, and you will have been compelled to supply the conclusion that "Socrates in mortal" on your own.

To understand how interfaces can persuade users as effectively as enthymemes, it's useful to consider Wolfgang Iser's concept of "constellations" from *The Act of Reading* (1978). Iser, who was a leading thinker in the reader response movement, claimed reading (and, by extension, viewing an interface) is an interactive activity rather than a passive act of consumption. The idea works as follows: Readers (or in our case, users) don't merely consume or absorb a story; instead, they fill in gaps in the narrative in the same way that people draw in the lines between stars in a constellation. By extension, I argue that users are engaged in the same gap-filling, constellation-building activity when confronted with interactions in an interface that requires they use *predictive power* to select an interaction.

Indeed, Iser's (1978) explanation of how readers create meaning by bridging gaps in a constellation resembles the idea of "conceptual models" interface that designers like Brenda Laurel use to describe the internal consistency of an interface's information architecture. In Chapter 2 of *The Computer as Theatre* (1993), Laurel explains how users are asked to use clues embedded in an application's interface to construct a conceptual model of the virtual world. This virtual world mirrors as closely as possible the conceptual model used by the information architects to create the software. Laurel, like Iser, thus suggests that users actively create and essentially write themselves into interpretive frameworks, or paradigmatic "constellations," to provide themselves with predictive power over an interface.

From the constructivist point of view, it is critical to develop this understanding of how users take advantage of social protocols (like handshaking) to fill gaps in an interface so they are interpellated into a conceptual model of the interface. This approach is essential in today's world of user-experience design because it tells usability professionals where to look when researching users for a product under development. It tells us that using our current usability research tools to assess whether or not products can accommodate users' needs and goals is no longer sufficient to meet our clients' needs. As Alan Cooper remarked in *The Inmates Are Running the Asylum* (2004), the result of accommodationist usability testing is a bit like using sandpaper when what you really need is a circular saw. The tool being used simply doesn't cut deep enough to reveal the root causes of the problems. Instead, we need to adapt our current research tools and develop new ones that will reveal the social, aesthetic, and cultural characteristics found in a particular user demographic. We need to help our clients understand what kinds of social protocols, aesthetic choices, and cultural cues they should use to interpellate users into subject positions that will enable them to predict and use the designer's conceptual model. By learning and using such tools, academics working in usability and user-centered design from a constructivist approach can more effectively help industry partners address a range of interests and problems.

One example of how the UTF worked for an industry partner to provide this kind support is in research we provided for a textbook publisher just starting the production of a new book for undergraduate public speaking courses. During focus group studies, even the teachers of the classes told the development team they really didn't know how students use public speaking textbooks outside of class to prepare speeches for class. Through initial research, the members of my team and I discovered that teachers and textbook authors really didn't know if students ignored the text entirely, if they located information in the index or table of contents and read small sections, or if they scanned or read whole chapters. The authors and development editors for the proposed book didn't know what examples students wanted, what tasks would drive students to the book for help, or how much information was too much and potentially insulting for the audience.

To answer these questions, we couldn't simply give students textbooks and conduct think-aloud protocols, particularly as we really didn't know what were realistic tasks students perform with the texts. We needed to create a hybrid study design, and we began by conducting a 6-week-long diary study. We asked 24 students enrolled in six different entry-level public speaking courses at both 2-year and 4-year institutions to send us diary entries twice each week. For these entries, students answered predetermined questions about what they were doing in their classes, whether they were using their textbooks in and outside of class, and if so, why were they using them, how, and for how long. We also collected information about whether the textbooks they were using met their needs and, if not, what they did to obtain needed information.

During the diary study, we conducted interviews with the students at the beginning, middle, and end of the 6 weeks. The research team also exchanged e-mail messages with individual participants throughout the period and asked them to further explain comments they made in their diary entries if entries were undeveloped, unclear, or surprising. To triangulate data collected from diaries, the research team conducted a brief think-aloud protocol analysis at the end, and it involved asking students to use their textbooks to perform tasks that focus groups and class observations revealed were typically expected in public speaking courses.

On average, students reported spending about one hour (62.2 minutes) per week outside of class studying for their public speaking course. Interestingly, students spent little time "studying" for class in the traditional sense. Most of the time that students devoted to public speaking outside of class involved researching speech topics and drafting speeches. Indeed, the results revealed that nearly 75% of students' time was spent on researching, drafting, or practicing their speeches. *Less than 15% of their time was spent reading the textbook.* So, on average, for the 62.2 total minutes students spent each week on their class, they spent only about 8.5 minutes of that time reading their textbooks.

The amount of time using their textbooks didn't vary much from institution to institution either. It was revealed that 2-year students reported spending nearly 40% less time studying compared to 4-year participants, and 4-year students also reported a higher percentage of time spent working online either looking for videos of sample speeches they could watch or collecting information on research that they could incorporate into their speeches. In general, however, there didn't seem to be a great deal of difference in the basic ways that the twelve 2-year students and the twelve 4-year students allocated their time.

In their diary entries, participants were asked to specify what material they read in their textbooks during the 6-week study, why they read it, and if they found it to be helpful. The findings indicate that students used their textbooks only when

- It was assigned because of a quiz or they were preparing for an exam
- Students were seeking examples of a particular type of speech
- Students were preparing outlines for a speech

- A reference work needed to be cited in the correct MLA format
- The instructor told them a section of the text would help them with a speech
- The instructor told them they would have to discuss the material in the next class
- Students were seeking practical advice on visuals/presentation-aid preparation
- Students had to write a critique of a speech and were seeking criteria to guide the analysis
- Students needed guidance with developing and designing group presentations

The overall sentiment from students was that their Public Speaking course was about *performance*. They felt that textbooks spend a lot of time telling them *about* public speaking, while the real priority for them was learning *how* to be an effective public speaker. All of the participants reported that most of their final grade was based on their speeches, and they felt tests and quizzes only represented 10%–20% of the total grade (their instructors reported that tests and quizzes on readings were 35%–40% of the total). Ultimately, students' fear of performing poorly in front of the class far outweighed their concern about performing poorly on a quiz or exam, and so they indicated dissatisfaction with textbooks that were focused on teaching them "about" public speaking, information which would appear in a quiz or exam. This perception was largely what motivated their strong desire for more practical how-to information and many, many more examples in a textbook.

At this point, it would be reasonable to ask, what is it about this study that makes it constructivist rather than accommodationist? After all, wouldn't the accommodationist approach want the same type of information? It is certainly the case that a lot of the data would be relevant if someone were asking to evaluate whether the textbooks being used in this study accommodated users' needs. But that wasn't the principal research question we were asking, for our client wasn't sure the company was going to produce a textbook, a DVD, a Web application, or some other type of information product. Indeed, our industry partner didn't have an author or content developer in mind, as it wasn't clear what approach would be needed for the product when we began the study.

Unlike the accommodationist studies we were doing in 1993, what makes this study different was that it was addressing different research questions—questions that revealed a significant split between the kinds of social protocols, aesthetic choices, and cultural cues that teachers expect in a public speaking textbook versus those that the student culture expects. As professionals with PhDs in communication studies, teachers of public speaking cared about the theories behind making speeches, and they wanted students to know "about" the field. As textbooks are largely written *by* teachers *for* teachers (who pick the texts their classes will use), it's not a surprise the focus of the books was on speaking the language of teachers. The data, however, revealed that the culture students inhabit is far different; their focus was on learning how to speak, not about speaking. So while much of the data are the same for an accommodationist, the focus here is on the users' *cultural concerns* and not just their evaluation of the product.

The second reason this project was a constructivist approach is that the final product didn't merely give users what they said they wanted. Because textbooks are selected for use by the teachers and not by the students who use them, no publisher can afford to design an information product that ignores teachers and wholly accommodates students. Instead, what we needed to do was to find ways to bridge the gap and to construct roles in the text that students could play that provided them with the interpretive framework they needed. We needed to understand the role that authors play in actually constructing the use or task environment for users.

Because of the emphasis on developing content (particularly in textbook publishing and information product development), our clients often overlook the role authors and designers play in the construction of users' task environment. From a constructivist point of view, successful texts and information products aren't merely about a "content developer" delivering accurate content clearly and concisely. To be truly successful in the constructivist approach, content developers must also create roles and provide interpretive frameworks that users can deploy to successfully complete tasks and achieve their goals for an information product. One way to do that is by using "If, then" scenarios that users can play. (See Flowers, Hayes, and Swartz article "Revising Functional Documents: The Scenario Principle" [1983] for more details about this approach).

We recommended that our textbook client use the scenario principle as a major structural element for delivering content in the book (e.g., using leading questions and statements like, "If you're conducting research for a speech and a source you are considering meets most of the X, Y, Z criteria, hold on to it."). We also recommended the whole text be based around the use of ministories, case studies, and anecdotes about problems other students faced, for these create a clear role for users to play as they use the text. Because the users understand the role they are supposed to play, they also understand the logic they're supposed to follow in order to make the decisions necessary to complete the task at hand.

In sum, we didn't merely document what was successful and what failed to meet students' needs in this study. Instead, what the UTF team recommended to our industry partner was a range of techniques (and because this was a sponsored study, I am able to reveal only a few of them here) the authors and developmental editors could use to create a textbook user experience that constructed the roles students would play as they used the text. UTF provided the client with personas, descriptions of the user demographic, detailed descriptions of the users' work environment, analyses of the users' background and education, a breakdown of the types of comments users made about competing texts, and much more information. (All of these items were to help our industry partners make informed decisions about how to design user experiences that *persuaded* users to engage in specific types of behaviors.)

SOME TAKE-AWAYS

If you're planning to start developing academic-industry partnerships based on usability testing and user-experience design services, what are some of the concepts to be learned from this history of the UTF? First, build on your strengths and seek senior leadership. The MAPC program was fortunate to have Dixie Goswami's experiences at the American Institute for Research to build on as we sought out corporate partners. Her knowledge of workplace cultures and industry needs was a major asset as our faculty reached out to industries in the community. But even if you can't develop an outreach program through your school's senior leadership, you can begin building it through faculty exchange programs and becoming involved in professional organizations like the STC's Usability and UX SIG.

Indeed, even with Dixie's leadership, I became involved in our local chapter of the STC so I could network with industry professionals. Using the resulting network, I developed a relationship with industry colleagues at AT&T, which allowed our MAPC faculty to set up an exchange program. I was able to work several months at AT&T, learning about their needs and workplace culture, while at the same time, their senior technical writers were able to teach writing classes for the university. Over time, this faculty exchange led to a long-term and multifaceted academic-industry partnership between the MAPC program and AT&T.

A second observation about the UTF's experiences is the success we had building on service-learning types of projects in our classes. I don't believe NCR's management would have been interested in funding sponsored graduate assistantships for the UTF if my colleague Susan Hilligoss and I hadn't first had our technical writing students working closely with NCR employees on writing projects. I have found that going directly to industry and asking for support to build a usability testing lab without first developing a relationship with them isn't an effective strategy. Having your students work on class projects with the company's developers, technical writers, engineers, or other employees is an excellent method of beginning to develop that relationship.

Third, be prepared to negotiate contracts with your industry partners so your reports protect the teams with whom you'll be working. This strategy can be difficult for many of us in academe, but you should be prepared to give up ownership of the research reports you provide your industry partners. You'll also need to be sensitive to their desire to control and even own the development processes for the products they're creating. Creating contracts that stipulate that you will communicate through a designated liaison or that limit the scope of work so that you aren't perceived as taking over from a development team is an important means of build sustainable relationships with industry partners.

Fourth, plan on being nimble and develop your team and infrastructure for flexibility. As I hope this retrospective has shown, even fields that seek to found themselves on conservative, stable empirical principles are still forced to change in the face of market pressures. What's more, every new study I've done, even though I may be working with an industry partner who has used the UTF for years, will require a new research design and a modified methodology. While we might be doing a think-aloud protocol for an established partner, the UTF team needs to constantly adapt and retool in order to meet the unique and particular needs of the research questions driving that specific product's design.

Finally, when I'm asked by colleagues about building a new usability testing lab at their school, I never recommend they invest significant funds in building a laboratory infrastructure. Too often, I've seen schools spend tens and even hundreds of thousands of dollars converting classrooms into lab facilities with one-way mirrors, closed-circuit cameras, microphones, and office computing environments. Indeed, many eye-tracking systems cost over $25,000 alone, not to mention the personnel costs involved in setting them up and training your staff to use them effectively. This isn't to say that you shouldn't invest in some infrastructure; it's certainly the case that I've had to buy or rent eye-trackers and other equipment necessary for research projects. Still, I fear the real cost of these infrastructure investments is the drag they place on methodological creativity. Too often I've seen a team of faculty researchers force a project into a predetermined research design because they have to justify the laboratory and equipment they acquired. It's human nature: If you have cool toys, then you want to use them. However, I believe that part of the UTF's longevity and success is that we have steadfastly refused to invest in data collection technologies until we could unequivocally demonstrate their usefulness over time, or unless we could easily throw them away once a study had been completed. Instead, we have chosen to invest in tools that allow us to stay nimble and focused on the research needs of our clients.

CONCLUSION

Today, developing academic-industry partnerships in the usability field is far more complex than it was 20 years ago, and most of the "low-hanging fruit" we were picking in the early 1990s is gone. There probably are a few companies still naïve enough about the usability of their products that they could benefit from an exclusively accommodationist approach to a partnership with them. I certainly still get potential clients who only want to have the UTF conduct an eye-tracking study that proves users are or are not looking at some desired location on the screen. The pendulum, however, has swung pretty far to the constructivist side of the equation these days, and clients are demanding a lot more specific information about the social protocols, aesthetic choices, and cultural cues their customers use with, not only their products but also those of their competitors.

More of our industry partners want an eye-tracking study that doesn't merely show where their customers are looking on the screen; they want actionable research that will enable them to persuade the user to look in a particular location. Thus, what I hope that this high-level overview of 20-years of academic-industry partnerships has shown is that identifying and keeping up with the evolving needs of clients' research needs is a necessary criterion for sustained success. It's certainly not a sufficient criterion, for many other factors are at play. Yet finding ways to communicate with potential clients about the field's evolution has been a major factor in the success of our partnerships.

REFERENCES

Anderson, S. P. (2012) *Seductive interaction design: Creating playful, fun, and effective user experiences.* Berkeley, CA: New Riders.

Althusser, L. (1971). Ideology and ideological state apparatuses (notes toward an investigation). In *Lenin and philosophy, and other essays* (B. Brewster, Trans.). New York, NY: Monthly Review.

Bogost, I. (2010). *Persuasive games: The expressive power of video games.* Cambridge, MA: MIT Press.

Burke, K. (1968). *Counter-statement.* Berkeley, CA: University of California Press.

Cialdini, R. B. (2006). *Influence: The psychology of persuasion.* New York, NY: Harper.

Cicero (1955). *Rhetorica ad herennium* (H. Caplan, Trans.). Cambridge, MA: Harvard University Press.

Cooper, A. (2004). *The inmates are running the asylum.* Indianapolis, IN: Sams.

Dumas, J., & Redish, J. (1993). *A practical guide to usability testing.* Westport, CT: Greenwood.

Fletcher, D. (2010, May 7). The 50 worst inventions: Microsoft Bob. *Time.* Retrieved from http://content.time.com/time/specials/packages/article/0,28804,1991915_1991909_1991855,00.html

Flower, L., Hayes, J., & Swartz, H. (1983). Revising functional documents: The scenario principle. In P. Anderson, R. Brockman, & C. Miller (Eds.), *New essays in technical and scientific communication: Research, theory, practice* (pp. 41–58). Amityville, NY: Baywood.

Fogg, B. J. (2002). *Persuasive technology: Using computers to change what we think and do.* San Francisco, CA: Morgan Kaufmann.

Grant-Davie, K. (1992). Coding data: Issues of validity, reliability, and interpretation. In G. Kirsch & P. Sullivan (Eds.), *Methods and methodology in composition research* (pp. 270–286). Carbondale, IL: Southern Illinois University Press.

Howard, T. W. (2010). *Design to thrive: Creating social networks and online communities that last.* San Francisco, CA: Morgan Kaufmann.

Iser, W. (1978). *The act of reading: A theory of aesthetic response.* Baltimore, MD: Johns Hopkins University Press.

Johnson, J. (2010). *Designing with the mind in mind: Simple guide to understanding user interface design rules* (1st ed.). San Francisco, CA: Morgan Kaufmann.

Laurel, B. (1993). *Computers as theatre*. Reading, MA: Addison-Wesley.

Molich, R. (1998). *Comparative evaluation of usability tests*. Paper presented at Usability Professionals Association conference, Washington, DC. Retrieved from http://www.dialogdesign.dk/tekster/cue1/cue1paper.pdf

Nielsen, J., & Mack, R. (1994). *Usability inspection methods*. New York, NY: John Wiley & Sons.

Norman, D. A. (2005). *Emotional design: Why we love (or hate) everyday things*. New York, NY: Basic.

Pink, D. (2011). *Drive: The surprising truth about what motivates us*. New York, NY: Riverhead Books.

Redish, J., & Barnum, C. (2011). Overlap, influence, intertwining: The interplay of UX and technical communication. *Journal of Usability Studies, 6*(3), 90–101.

van Gorp, T., & Adam, E. (2012). *Design for emotion*. San Francisco, CA: Morgan Kauffman.

Weinschenk, S. M. (2009). *Neuro web design: What makes them click?* Berkeley, CA: New Riders.

Weinschenk, S. (2012). *100 things every designer needs to know about people*. Berkeley, CA: New Riders.

CHAPTER 9

Making Space for Community Voices: Rhetoric, Engagement, and the Possibilities for Partnerships

James M. Dubinsky

> We came with vision but not with sight. We came with visions of former places but not the sight to see where we are.
>
> —Wendell Berry, *The Unsettling of America* (1977)

ABSTRACT

My chapter features a case study of the Community Voices (CV) project, a regional, cross-sectoral collaboration that involves an interdisciplinary team of teachers and students. The CV project provides a local forum for sharing diverse perspectives on community resilience and innovation to build community capacity and encourage creative leadership. It also seeks to advance its value proposition to the university and to the New River Valley community by developing creative dialogue strategies effective in contexts of dispute and conflict. The case analysis presented here affirms characteristics of effective partnerships, demonstrates the importance of understanding the power of "place," and exemplifies how academy-industry partnerships—centered on an engagement approach—can help our students and colleagues meet the technical, social, and human challenges our world faces.

INTRODUCTION

In 2007, I was asked to serve as co-chair of a university task force charged with identifying links between student engagement with the community and the university's land-grant mission. The task force outlined a multiyear plan predicated upon the creation of a new university center, the Center for Student Engagement and Community Partnerships (CSECP). (The center is now called VT-ENGAGE.) I was invited to establish that center as its founding director, and my 4-year involvement in that project helped me continue a journey to understand the roles for and importance of engagement, particularly civic engagement, in the lives of students and faculty.

During that journey, I learned much from scholars such as Robert Greenleaf (1972), who called for higher education to address pressing social issues and needs:

> If a better society is to be built, one that is more just and more loving, one that provides greater creative opportunity for its people, then the most open course is to raise both the capacity to serve and the very performance as servant of existing major institutions by new regenerative forces operating within them. (p. 1)

Greenleaf's credo, that organizations have the ability to raise the capacity to serve, provides a clear rationale for civic engagement in higher education.

In this chapter, I move from individual classrooms and programs toward larger, institutional partnerships, demonstrating how Community Voices (CV), a partnership based upon Greenleaf's foundation of servant leadership, offers benefits in terms of regeneration, creativity, and problem solving, all the while moving ideas of engagement forward. My thesis that CV project exemplifies ways academy-industry partnerships, centered on an engagement approach, can help our students and colleagues meet the technical, social, and human challenges our world faces.

ASSETS SOLVE PROBLEMS: PROBLEM STATEMENTS JUST NAME THEM

Greenleaf's (1972) concept of "servant leadership" is part of a long tradition going back to ancient China, Greece, and Rome, and it is a tradition that has accelerated in the past 25 years with the advent of Campus Compact in 1985, the establishment of the Carnegie Community Engagement Elective Classification designation in 2005 (see Carnegie, 2012), and the creation of the Engagement Academy for University Leaders sponsored by Virginia Tech's Center for Organizational and Technological Advancement in 2008 (see COTA, 2012). All of these efforts are linked to responses from university and colleges to public criticism about not living up to their promises to society, responses that were codified with the *Presidents' Fourth of July Declaration on the Civic Responsibility of Higher Education* (Campus Compact, 1999).

In this *Declaration*, university leaders explained that they had "a fundamental task to renew [their] role as agents of democracy" and to reinvigorate the civic role of higher education. The underlying premise, based upon over two decades of work with action research and service learning, was that student learning would be enhanced through carefully structured curricular and co-curricular activities involving community service and action research (Sirianni & Friedland, 2005). However, while the presidents' goal was and remains laudable, "putting knowledge to work" is demanding and difficult. The fundamental challenges of such work include creating a bridge between campus and community—a bridge based on both trust and relationships that allows campus representatives to gain a clear sense of the local community's needs and that allows them to resolve issues concerning faculty roles and rewards and resource allocation.

Usually, when addressing challenges of this or any type, many individuals choose to begin with a problem statement. Here is one such statement I wrote as part of a grant proposal:

> In addition to a series of traumas affecting our university community,[1] there has also been the powerful, social impact of the economic downturn in our local and extended community. The unemployment rate in most of the local economic district (New River Valley, which consists of four counties and the city of Radford) has been consistently higher than Virginia state averages over the past ten years and some localities (Pulaski and Giles Counties) have been higher than national averages (Virginia Employment Commission). Evaluating a 15-month trend in unemployment rates reveals a steady incline in unemployment, which for most of the area is higher than state and national averages. The loss of jobs; the halting or reversal, in some cases, of income; and the powerful, emotional impact of the traumas have combined to place tremendous stress on our local communities.
>
> Sadly, what our community is facing has been described by some researchers as the "new normal," a period where, even though the initial shock and distress levels may subside, "there remains persistent, widespread distress that . . . is significant in terms of social functioning and adaptation." (Steury, Spencer, & Parkinson, 2004, p. 158, as cited in Dubinsky, 2009)

When applying for grant money, such a strategy has its advantages. To receive funding, organizations must offer a justification that relates to the mission of the funder. However, when granting agencies are not involved, and organizations are trying to solve problems such as those described above by relying on local

[1] On April 16, 2007, Virginia Tech experienced one of the worst on-campus shooting incidents in history: 32 students were killed, 58 were wounded, and thousands of people were traumatized.

resources, such a strategy might not be the most effective way to proceed. When faced with seemingly intractable problems, a broader, holistic strategy is needed, one that focuses on available resources, on nurturing mutually beneficial relationships with potential partners, and on developing an understanding of the power of "place."

A focus on place helps reduce what philosopher Jon Wagner (1993) argues are "specific areas of ignorance," which are often the result of a lack of context. Place knowledge is essential for engagement efforts to discover the necessary practical knowledge to address the needs present. In *Becoming Native to this Place* (1996), Wes Jackson, founder of the Land Institute and a respected advocate for sustainable practices, challenges universities "to stop and think what they are doing with the young men and women they are supposed to be preparing for the future" (p. 3). Jackson's point is that sustainability is not possible until all of us can become "native to our places in a coherent community that is in turn embedded in the ecological realities of its surrounding landscape" (p. 3).

Many educators have addressed this challenge. Some, such as Tal Stanley (2007), focus explicitly on the critical importance of understanding place:

> Whether the work is curricular-based service or extra-curricular, all of the service initiated . . . adheres to several standing rules. The . . . partnerships always seek long-term, sustainable solutions to problems and have a goal of helping to build a democratic future for all people of the place. (p. 311)

Others, such as Waters, Haninen, and Hardin (2011), focus on place less directly, highlighting the importance of understanding the needs of local agencies, and in so doing, getting "buy-in" (p. 73). Whether the focus or emphasis is direct or less so, the process requires "time and attention from the administering university" (Waters et al., 2011, p. 73). I would add that such time and attention are critical from a department or faculty member's perspective. Investing time in seeking knowledge about the local context, while recognizing how we are often viewed with suspicion, will improve the chances for a true, reciprocal partnership.

The Community Voices (CV) project is predicated upon a deep understanding of place, as well as a desire to build respectful partnerships focusing on sustainable solutions to endemic problems. The CV project provides a forum for all citizens of our region to share diverse perspectives on issues such as community resilience and innovative ways to build community capacity and encourage creative leadership. It is a step in helping the local community understand its problems and recognize some of its assets available, in terms of leaders who might be able to define and solve problems.

The goals of the CV leadership team include facilitating dialogue on topics determined by community participants themselves. The team also works to negotiate differences in values and worldviews among many community stakeholders with the long-term goal of creating sustainable social change. Sustainable change requires a different perspective, one that begins with an evaluation of assets, such as availability of higher education and two regional airports (New River Planning District, 2009, p. 21), and those assets yet to be identified.

In *Building Communities from the Inside Out* (1993), John Kretzmann and John McKnight focus on this strategy of "beginning with a clear commitment to discovering a community's capacities and assets" (p. 1). This approach seems sound to me as someone who has been a chair of a board of a large nonprofit serving a rural community with a fairly high unemployment rate and a relatively low average family income. However, to many individuals, this approach seems counterintuitive. In most situations, the strategy people begin with is similar to the one highlighted earlier for grant proposals: "needs- or problem-driven."

Unfortunately, with a problem-driven focus, there is seldom just one problem. When individuals say they wish to address the problem of affordable health care, for example, they often overlook the various factors (or component problems) that contribute to this overall situation. So then the dilemma becomes "Where do we start? How do we sort out a plan of action?" Such a dilemma can lead to paralysis or an inability to build any kind of clear consensus as to direction. Even if the group/organization can focus in on a problem, the focus becomes narrow, leaving the remaining issues to take up much if not most of the members' mental, physical, and most important, emotional real estate.

Another reason this approach is often inadequate is that by focusing on problems, we tend to look outside of our organization for solutions and for experts who can help. Keeping this somewhat paradoxical situation in mind can make a difference when trying to create cross-sectoral partnerships, such as those between universities and community organizations. In many cases, faculty, and especially students, see themselves (or hope they are seen) as experts who can solve the problems. Often they are such experts, but the expert/nonexpert frame can create hierarchies and/or dependency and usually does not lead to sustainable solutions.

Several years ago, Bob Siegle (2005), a colleague selected for one of our university's teacher-scholar awards, explained that

- Teachers are not the experts.
- Students are not the ignorant.
- The answer to *what* to do lies in the *form* of the doing.

This mindset applies equally when walking into the community: We should walk in not as "experts," but as participants in a learning process.

If, rather than beginning a partnership with the mindset of the expert problem-solver, those of us from universities would begin as participants who want to help document a community's strengths, gifts, and talents, partnerships are more likely to thrive and be sustainable. As Hauser (2004) states

> Civil society's . . . web of associations are rhetorical arenas in which "strangers encounter difference, learn of the other's interests, develop understanding of where there are common goals, and where they may develop the levels of trust necessary for them to function in a world of mutual dependency." (p. 9)

These strengths and associations can be mobilized or combined into teams that have a capacity for action and innovation. So, in reviewing assets, we focus on what we have and start to make connections. These connections lead to what Siriani and Friedland (2005) call "civic renewal"; what Levine (2007) labels "the future of democracy"; and what Harry Boyte (2008) refers to as the "citizen solution": how to get "beyond slash-and-burn politics" and create "effective civic action" that can solve public problems (Boyte, 2008, p. 3).

As scholars and community advocates, members of the CV team recognize that community leaders who have been working to address the issues and problems native to our area are its most important assets. By seeking those leaders' voices, the team seeks to understand the dynamics of the processes involved in negotiating differences in values and worldviews among many community stakeholders and study whether certain strategies work more effectively than others to create change. As educators, we hope to have a positive impact on the next generation of citizens' belief in democracy as a process involving people in communities who cooperate, invest, and commit themselves to achieving collaborative solutions that benefit the widest range of people. We advocate an action-research model that fosters informed, structured dialogue (Freire, 1993; Habermas, 1984).

OUR REGION AND ITS ISSUES

Perhaps one of the most important elements of any action-research model connecting the academy and industry is knowing where to start. In individual classes, once a specific partner has been identified, a sound first step is to ask students to create a frame, starting with some background knowledge. That frame can be part of a written project (e.g., a report on the history, mission, and values of the organization or community being considered as a partner).

In Community Voice's case, we began by seeking such foundational data for our rural region. In that research, we outlined how our region is beset with specific problems, including long-term unemployment and loss of manufacturing jobs, issues of identity, and the need to establish a common vision for the future (New River Valley Planning District Commission, 2009). Because we are blessed with natural assets such as a strong community foundation and two major universities, our region has hope for "innovation . . . and increasing diversity" (Richardson, 2009, p. 44). The complication is that our region is changing: Once deeply rooted in agriculture but now more recently home to many high-tech businesses, our region has become a place of contentious debate and serious questions about the place of citizenship and the citizenship of place, as well as the value/relevance of compromise. The larger problem is a lack of trust and an inability to talk openly and civilly about specific problems.

A recent series of events concerning a $1million grant from the Federal Partnership for Sustainable Communities, administered by the NRV Planning District Commission (NRVPDC), illustrates these contentious debates. The project invites a consortium of local governments, professional and business organizations, nonprofits, state agencies, and educational institutions representing the entire region to collaborate with the broader public on a common vision for our region's future. The result of this 3-year project will be a 20+ year *Livability Plan* to propose policy options that would address the region's current and emerging needs regarding housing, transportation, energy, agricultural systems, natural resources, technology infrastructure, economic and workforce development, water resources, and cultural assets.

Although the initiative is in the initial stages of recruiting partners and developing a strategy for broad public participation, there has been significant opposition by members of the local Tea Party and other politically conservative leaders. Such opposition is based upon problems similar to those in other rural areas across our nation: displacement of workers from the manufacturing sector, a shortage of middle-class jobs with benefits, problems implementing government grants that positively affect the widely dispersed populations' needs (e.g., good roads, better public transportation), not to mention a general distrust of the national government.

The ability to dialogue with one's neighbors who have differing points of view has long been the lifeblood of democracy and something that people in rural regions, who have shared long histories together, found easy to do until recently. The tenor of this debate, particularly at recent meetings of the Pulaski County (Virginia) Board of Supervisors, suggests that our ability as neighbors to engage in civil dialogue, respectfully disagree, and come to a compromise is waning. Instead, the debate has been characterized by thin evidence, calls to unseat political leaders, and a general unwillingness by parties on all sides to attempt to understand opposing views and move forward in a positive direction. For example, our region's newspaper, *The Roanoke Times,* reported that "Hundreds turned out [for what] at times became charged with emotion, from cheers to jeers to name-calling." At least one citizen took an extreme position, claiming the proposed plan would "ruin this country." This same person later compared the (livability/sustainability) initiative with the removal of Jews from their homes during the Holocaust (Matze-Fawcett, 2011).

Other attendees noted similar conflicts and tensions. A local blogger (Hayne, 2011) with 48 years of residence explained,

> As we moved further into the discussion, it became them and us. The no government five vetoed all but their own suggestions: they wanted a future

> which insured property rights, with no gov't intervention, a self-reliant self-sustainable valley. They wanted jobs, but they failed to offer any ideas for attracting them. (Matze-Fawcett, 2011 {TH})

With such dramatic comparisons and nicknames that have a partisan flavor ("no government five"), not unlike those made by national leaders in the debates over health care, gun control, or the budget, little hope of getting beyond sound bites or reaching collaborative solutions is possible.

The erosion of deliberative dialogue in our public institutions and community centers such as churches, libraries, and civic organizations portends an uncertain future for our communities and sense of responsibility as citizens. The CV project's model provides critical spaces to learn from and identify with local and regional community leaders who have successfully navigated similar contexts. The leadership team dialogues with these leaders and others in the area and then tests strategies to create civil, deliberative dialogue. In so doing, the team identifies leaders with "cultural fluency skills"—the ability to develop processes to transcend boundaries, listen well, work well with dissent and difference, and solve problems (Eng, 2009).

Such a model is transferrable to the classroom as well. As outlined above, the focus is on understanding a range of issues fostered by organizational cultures and working to build bridges between or among them. One way is to start with some short- and long-term visioning with the community partner that focuses on self-interests and limited goals (Sirianni & Friedland, 2005, p. 140). Study circles, public issues forums, or the World Café model (Brown & Isaacs, 2005) all are approaches that can help create a dialogic space for such visioning. All involve moving beyond the walls of the classroom as well as demonstrating that the classroom is available and accessible to community members. This kind of community building and narrative inquiry demonstrate key characteristics of modern action research: "its participatory character, its democratic impulse, and its simultaneous contribution to social science and social change" (Carr & Kemmis, 1986, pp. 163–164).

A WORKING INITIATIVE

When preparing for this kind of community organizing and development, another good strategy is to find best practices within the community, as they will almost always have taken into account many of the local problems based in power relations because they will have incorporated a focus on "place." One initiative in our region, led by SustainFloyd (SF), a grassroots social benefit organization, is providing innovative leadership to build a sustainable regional community along the Blue Ridge Plateau (Floyd, Carroll, and Grayson Counties). SF's concept of smart change consists of several steps: variation, survivability, and selection. Having developed a working partnership with a base of local donors, local government, Virginia Tech, the Appalachian Regional Commission, the USDA, and the Virginia Tobacco Commission, SustainFloyd has completed a local energy inventory, established a community farmers and artisans market, and secured funding for feasibility studies for community food and dairy processing facilities—all in 18 months.

Having established credibility in the community by taking tangible actions responsive to local needs, SustainFloyd is turning to the task of creating dialogue and finding common ground with the broader community. Recent expressions of outspoken opposition to sustainable development initiatives across the New River Valley create problems for progressive organizations such as SF. One outcome of the CV team's work is to assist organizations such as SustainFloyd to serve as arbiters and/or bridge builders.

Many strategies for building partnerships in other communities have been documented. In a qualitative study of 99 community partners across eight California communities, Sandy and Holland (2006) found several strategies that worked, including the need for "partners to engage in ongoing relationship-building" (p. 39) and holding "conversations regularly about [the] partnership process and outcomes" (p. 40). Creating partnerships that last and are sustainable requires time, sensitivity to distinct needs, and a willingness to invest in creating networks. When examining all of them, at the core is a need to be able to talk to one another in an open, dialogic, and respectful way. Creating such dialogue has been at the core of the CV project, and it must be at the core of any individual classroom or program-related project.

CV SERIES—DIALOGIC WORK

Between January and May of 2011, the CV project hosted its first three events, featuring community leaders from the public, nonprofit, and private sectors as speakers: Anthony Flaccavento, Dudley Cocke, and Beth Obenshain. Prior to the events, members of the team interviewed the speakers and later transcribed those interviews. During the events, the speakers shared their personal and organizational stories and insights with specific reference to the designated themes and a focus on new and actionable ideas and strategies to support community change and development. Their presentations and the discussions that followed formed the first part of Community Voices' comprehensive and focused program to generate communitywide dialogue around a designated common theme: the creative, imaginative, and ethical dimensions of leadership for community resilience and innovation.

During the 2011–2012 academic year, the Community Voices project convened six additional CV events with the same structure. The goal remained focused on uniting the various constituents in the region by having community leaders articulate how addressing the issues and solving local problems require collaboration, imagination, and a mobilization of the community's assets, most specifically, its citizenry. The way "to" is *through*, and that way is led by "fluent leaders" who "understand the value of nuance" and who are capable of dealing with "complexity, knowing that solutions to real problems require many levels of information and analyses from many sources" (Eng, 2009, p. 35). Such fluent leaders offer their personal interpretations of how change is possible amidst significant differences, often the result of deeply held values and therefore different epistemic perspectives.

The speakers for the next CV series had, in the course of their professional lives, proven themselves "fluent." They ranged from regional civic leaders (e.g., Nancy Agee, the CEO of Carilion Hospital) to international leaders, such as Dr. Avila Kilmurray, who directs the Community Foundation of Northern Ireland and has written widely on creating community dialogue. The CV leadership team believes these speakers provide insights about the issue of how leaders/leadership emerges within context. These themes are linked to the issue of livability, which is, as pointed out earlier, a concept in contention in the local region.

BUILDING BRIDGES AND RELYING ON COMMUNITY ASSETS

Over the past several years, the Community Voices leadership team has been researching and working to more fully map the community's assets, its identity, and its strategic, long-term goals. The team has involved stakeholders in the discussion/visioning process and relied upon proven methods for involving citizens (Bowie, Richardson, & Sykes, 1995; Fishkin, 1997; Morrison, 2003; Stott, 2009). The result has been a strategic plan of sorts, a plan that fosters, in Peter Senge's (1990) words, a learning organization "where people continually expand their capacity to create the results they truly desire, where new and expansive patterns of thinking are nurtured, where collective aspiration is set free, and where people are continually learning how to learn together" (p. 3) or what some, such as Sanders (2001) and Inakaga (2007) have labeled "collective creativity." The benefits of such creativity include learning to

- Think together in teams;
- Listen actively;
- Share with a wide range of people in small, intimate settings.

Creating a dialogic, collective creativity counters a powerful force in our culture, one labeled by scholars such as Bellah, Madsen, Sullivan, Swidler, and Tipton (1996) and Robert Putnam (2000) as "radical individualism." Such a force should not seem hard to understand; the power of the individual has deep roots in our country's mythology: "I'll do it my way"; "Go your own way"; "You go your way, and I'll go mine." The end result is an organization in which the majority of its members practice and are committed to what Benjamin Barber

(1984) calls "weak democracy." This kind of democracy produces "citizens" whose only obligation is to vote and pay taxes so that others can provide for their needs: build their roads, tend their sick and elderly, and defend their shores. The result? A country, or in our case an organization, that focuses on "busyness and consumerism," which might "diminish and harden our sense of what it means to be human" (Daloz, 1996, p. 11).

But not everyone believes in weak democracy. Some scholars, such as Barber (1994), advocate a strong democracy, a society predicated upon inquiry, deliberative discourse, and education. This education, similar to the one practiced in ancient Athens, "equips a citizen for genuine choice of a way of life, . . . requires active inquiry, and the ability to contrast alternatives" (Nussbaum, 1998, p. 54). With this kind of education, students become citizens who recognize

> what is especially fundamental about [people], most worthy of reverence and acknowledgement, namely their aspirations to justice and goodness and their capacities for reasoning. It creates a series of concentric circles and we learn about humanity as a whole, while not giving up our special affections or identifications; we learn to understand differences. (Nussbaum, 1998, p. 60)

The kind of engagement brought about by servant leadership and developed in partnerships that are mutually beneficial and focused on the strengths and assets of the community will foster this kind of citizenship.

By working together—understanding in more intimate detail about how all of our students and community members live and work—what we learn out there becomes part of who we are *in here*. It becomes part of *us*, and it gives us some new assumptions or attitudes. These assumptions and attitudes are the "infrastructure upon which our community rests," one that is "situated under our feet and in our heads" (Calorusso, 2007, p. 108). And "it is precisely these discursive realities, carried in [our] heads . . . which can ultimately determine the success or failure of any . . . project" (Calorusso, 2007, p. 108). Maintaining or trying to maintain this infrastructure is integral to being native to a place.

These discursive realities and this infrastructure are under siege by both technological and social forces. In *You are Not a Gadget* (2010), Jaron Lanier, a Scholar in Residence at Microsoft and at the Interdisciplinary Center for Entrepreneurship and Technology at Berkeley (and the person who coined the term "virtual reality"), argues that due to certain kinds of protocols, most "words" will be

> read by nonpersons—automatons or numb mobs composed of people who are no longer acting as individuals. . . . [these words] will be scanned, rehashed, and misrepresented by crowds of quick and sloppy readers into wikis and automatically aggregated wireless text message streams. . . . The fates of these words will take place almost entirely in the lifeless world of pure information. Real human eyes will read these words in only a minority of cases. (p. ix)

That potential problem, based upon a deep understanding of the limitations of computing, worries him deeply, which is the reason for his book.

Lanier's solution? "You have to be somebody before you can share yourself" (2010, p. xiii). Ending there, without any further elaboration, leads the reader to speculate on the logical claim(s) embedded or entailed in that statement. Clearly, the ability to "share oneself" has some value or importance. Share yourself with "whom" and why are not addressed; in some sense, just the overall goal of deemphasizing the individual and emphasizing the collective. In asking/hoping for or encouraging sharing, Lanier is asking people to be people, a word that in and of itself points to a collective. The collective has its less-than-ideal qualities, one of the worst being when the individual is deemphasized and people revert to mob-like behavior (p. 19).

That mob-like behavior can, however, be modified or ameliorated when attempts are made to make large and powerful institutions, such as a major research university, local and friendly. We believe the model of Community Voices is one such method of transformation. By shifting from a top-down, problem-solving, expert mentality to one rooted in dialogue and mutual respect, our institution morphs from an organization that might choose to impose or create a "mob-like" mentality into an integrated part of the community, linked to the region's roots, with a central focus on human storytelling.

CONCLUSION

As I have explained elsewhere (Dubinsky, 2010b), providing direction, offering opportunities, and outlining a way—a *land-grant way*—for learning and serving, for engagement, is a vehicle to help students and community members understand the power of public or civic engagement as part of higher education's mission. Making it explicit that land-grant universities such as Virginia Tech were founded to educate students in the practical arts of agriculture and engineering and to serve the public directly through outreach and extension activities helps community members see past the institution's walls and begin to imagine it as an integrated part of their landscape, as "native." As a result of that explicit discussion, students, faculty, and community members learn that community/university partnerships extend and/or dissolve the walls and boundaries of the university campus, providing rich, theoretical settings for what scholars have called "connected knowing" (Schultz, 1987). The roots of this type of knowing go deep into the concept of place. Providing such a partnership and connected knowing helps to preserve memory and local knowledge because local stories are valued (Berry, 1990, p. 157).

Yet, even with the best of circumstances and the most open, explicit conversation about intention, there are risks involved. These risks, what I call the three "R's of service-learning"—Risk, Reciprocity, and Reflection—often are overlooked at the potential expense of entire projects and partnerships. As Carrick, Himley, and Jacobi (2000) have said,

> We have to remain alert to the power asymmetries and different discursive and material realities of the people involved in community-based projects. We risk confusing our ethical and political desires for reciprocal and mutually beneficial relations with the much messier realities that those relations often (re) enact. We risk masking rather than unmasking power dynamics. We risk mis-recognizing our own desires and needs. If we move too quickly . . . , we risk fixing complexities rather than acknowledging them as central to and part of learning. (p. 301)

Thus, all of us in Community Voices have learned to plan carefully and work to build partnerships. To achieve reciprocity, the second "R," requires dialogue and time spent listening.

To create a culture of engagement rather than one of service, especially avoiding service that is or can be viewed as simply "gift-giving" or "charity," requires a level of explicit discussion, project planning, and team building (Sibbert, 2002). Those working to "reach out" have to be explicit in providing direction, offering opportunities, and outlining our process for learning and serving.

With reflection—the third "R"—everyone involved in the partnership can bridge knowledge and experience. With university-community partnerships as vehicles for engagement, for going public, the binaries of theory and practice—form and content—are no longer valid. Instead, the process of teaching and learning becomes a möbius strip that turns constantly in on itself, wherein theory becomes practice, teaching becomes research. It becomes what Aristotle calls *techne*, a productive art, one in which, as Frost (1969, p. 275), put it, "love and need are one,/And the work is play for mortal stakes." I can think of no more powerful work than living a life undivided, focusing on collaboration in the classroom and community, willing to take risks with students to explore, contemplate, and co-create through vibrant, living, complex partnerships.

Building linkages between traditional learning and lived experience has been one of the primary tenets of the CV project. Working with partners to bridge knowledge and experience through an understanding of "place" isn't easy; it takes time and a willingness to reflect upon successes and failures. It requires the willingness to serve and a commitment to lead, without a desire to earn immediate or even long-term rewards. Service is itself a challenge. Robert Coles (1994) calls it a "rationale for life." With this knowledge, we have a means of understanding who we are and what we are capable of doing using the building blocks of language, the building blocks of a community of teacher-scholars in harmony with the others in our local communities. In Lanier's (2010) terms, we share ourselves, "create shared experiences" and "escape boundaries of pre-existing symbol systems" (p. 191).

Nearly every educator has heard of the 3 "Rs"—"Reading, wRiting, and aRithmatic." But many have not heard about the 3 "Rs" of service-learning or engagement—Risk, Reciprocity, and Reflection—with *risk* being front and center. What I am proposing in this chapter is neither easy nor necessarily a panacea; however, I believe, if done carefully and with eyes wide open, the strategy of shifting from teaching to learning, from service to engagement offers an opportunity for *work that is real*. My hope is that this chapter may open a door or a window into some possible ways to consider the möbius strip of teaching and learning in a world facing challenges linked to issues of ethics, privacy, and corporate and civic responsibility, into a world in crisis with so many people in need.

REFERENCES

Barber, B. R. (1984). *Strong democracy:Participatory politics for a new age*. Berkeley, CA: University of California Press.

Bellah, R. N., Madsen, R., Sullivan, W. M., Swidler, A., & Tipton, S. M. (1996). *Habits of the heart: Individualism and commitment in American life*. Berkeley, CA: University of California Press.

Berry, W. (1977). *The unsettling of America*. San Francisco, CA: Sierra Book Club.

Berry, W. (1990). The work of local culture. In W. Berry (Ed.), *What are people for?* (pp. 153–169). Berkeley, CA: North Point.

Bowie, C., Richardson, A., & Sykes, W. (1995). Consulting the public about health service priorities. *British Medical Journal, 311,* 1155–1158.

Boyte, H. (2008). *The citizen solution: How you can make a difference*. St. Paul, MN: Minnesota Historical Society Press.

Brown, J., & Isaacs, D. (2005). *The World Café: Shaping our futures through conversations that matter*. San Francisco, CA: Berrett-Kohler.

Calorusso, C. (2007). Understanding place as discursive space: A role for discourse analysis in the public design process. *Proceedings of Negotiating Landscapes conference*, CELA (Council of Educators in Landscape Architecture) (pp. 108–117).

Campus Compact. (1999). *Presidents' declaration on the civic responsibility of higher education.* Retrieved September 20, 2012 from http://www.compact.org/resources-for-presidents/presidents-declaration-on-the-civic-responsibility-of-higher-edu/

Campus Compact. (2012). *Who we are.* Retrieved October 10, 2012, from http://www.campuscompact.org/about/history-mission-vision/?

Carnegie Foundation for the Advancement of Teaching. (2012). *The Carnegie classification of institutions of higher education.* Retrieved October 12, 2012, from http://classifications.carnegiefoundation.org/

Carr, W., & Kemmis, S. (1986). *Becoming critical: Education, knowledge, and action research*. Philadelphia, PA: Falmer.

Carrick, T. H., Himley, M., & Jacobi, T. (2000). Ruptura: Acknowledging the lost subjects of the service learning story. *Language and Learning Across the Disciplines, 4*(3), 56–75.

Center for Organizational and Technological Advancement (COTA). (2012). *Engagement academy for leaders*. Retrieved October 12, 2012, from http://www2.cota.vt.edu/engagementacademy/index.html

Coles, R. (1994). *The call of service*. Boston, MA: Houghton-Mifflin.

Daloz, L. A. (Ed.). (1996). *Common fire: Lives of commitment in a complex world*. Boston, MA: Beacon.

Dubinsky, J. (2002). Service-learning as a path to virtue: The ideal orator in professional communication. *Michigan Journal of Community Service Learning, 8*(3), 61–74.

Dubinsky, J. (2009). *Restoring community: Strengthening networks of support for Virginia Tech and the New River Valley*. Grant proposal to the Dart Foundation.

Dubinsky, J. (2010a). A techne for citizens. In D. Franke, A. Reid, & A. Di Renzo (Eds.), *Design discourse: Composing and revising programs in professional and technical writing* (pp. 277–296). West Lafayette, IN: Parlor.

Dubinsky, J. (2010b). The land-grant way—Connected knowing and the call of service. *Present Tense, 1*(1). Retrieved October 10, 2012, from: http://www.presenttense journal.org/

Eng, P. (2009). The era of the bridge builder: Identifying the qualities of "fluent leaders." *National Civic Review, 98*(3), 34–39.

Fishkin, J. (1997). *The voice of the people: Public opinion and democracy*. New Haven, CT: Yale University Press.

Freire, P. (1993). *Education for critical consciousness*. M. B. Ramos (Trans.). New York, NY: Seabury.

Frost, R. (1969). *The poetry of Robert Frost: The collected poems, complete and unabridged*. E. C. Lathem (Ed.). New York, NY: Holt.

Greenleaf, R. (1972). *The institution as servant*. Terre Haute, IN: Center for Applied Studies.

Habermas, J. (1984). *Reason and the rationalization of society. Vol 1., Theory of communicative action*. Boston, MA: Beacon.

Hauser, G. (2004). *Rhetorical democracy and civic engagement*. In G. Hauser & A. Grim (Eds.), *Rhetorical democracy* (pp. 1–14). Mahwah, NJ: Lawrence Erlbaum.

Hayne, M. (2011, August 12). Livability initiative. *Tic toc tic toc blog*. Retrieved October 12, 2012, from http://tictochollar.blogspot.com/2011/08/livability-initiative.html

Inakaga, M. (2007). Collective creativity: Toward a new paradigm for creative culture. *Proceedings of the 2nd international conference on Digital Interactive Media in Entertainment and Arts*. Retrieved October 12, 2012, from http://dl.acm.org/citation.cfm?id=1306822

Jackson, W. (1996). *Becoming native to this place*. Berkeley, CA: Counterpoint.

Kretzmann, J., & McKnight, J. (1993). *Building communities from the inside out: A path toward finding and mobilizing a community's assets*. Chicago, IL: ACTA P.

Lanier, J. (2010). *You are not a gadget*. New York, NY: Knopf.

Levine, P. (2007). *The future of democracy: Developing the next generation of American citizens*. Medford, MA/Hanover, NH: Tufts University Press/University Press of New England.

Matze-Fawcett, A. (2011, July 26). Livability initiative draws ire in Pulaski County. *The Roanoke Times*. Retrieved November 2, 2012, from http://ww2.roanoke.com/news/nrv/wb/294112/

Morrison, D. E. (2003). Good and bad practice in focus group research. In V. Nightingale & K. Ross (Eds.), *Critical readings: Media and audiences* (pp. 59–84). Maidenhead, UK: Open University Press.

New River Valley Planning District Commission. (2009). *Community first: Regional partners in a global marketplace.* Retrieved November 2, 2012, from http://www.nrvpdc.org/CEDSVision2020report6-30-09.pdf

Nussbaum, M. (1998). *Cultivating humanity: A classical defense of reform in liberal education.* Cambridge, MA: Harvard University Press.

Putnam, R. D. (2000). *Bowling alone: The collapse and revival of American community.* New York, NY: Simon & Schuster.

Richardson, J. A., Jr. (2009). Fulfilling the promise: Rural America and the audacity of hope. *National Civic Review, 98*(3), 43–45.

Sanders, L. (2001, August). Collective creativity. *LOOP: AIGA Journal of Interaction Design Education, 3,* 1–6.

Sandy, M., & Holland, B. (2006). Different worlds and common ground: Community partner perspectives on campus-community partnerships. *Michigan Journal of Community Service Learning, 13*(1), 30–43.

Schultz, S. K. (1987). *Learning by heart: The role of action in civic education.* Paper presented at the 16th annual conference of the National Society for Internships and Experiential Education, Smuggler's Notch, VT.

Senge, P. (1990). *The fifth discipline.* New York, NY: Doubleday/Currency.

Sibbet, D. (2003). *Principles of facilitation: The purpose and potential of leading group process.* San Francisco, CA: Grove Consultants International.

Siegle, R. (2005). *Address to Diggs teaching scholar roundtable.* Virginia Tech, October 2005.

Sirianni, C., & Friedland, L. (2005). *The civic renewal movement: Community building and democracy in the U.S.* Dayton, OH: Kettering Foundation.

Stanley, T. A. (2007). Renaming the world in the valley of the Holston. *Appalachian Journal, 304*(¾), 302–315.

Steury, S., Spencer, S., & Parkinson, G. W. (2004). The social context of recovery. *Psychiatry, 67*(2), 158–163.

Stott, A. (2009). Citizen engagement. *Intergovernmental Solutions Newsletter Fall 2009,* 5–6. Retrieved November 2, 2012, from http://www.usaservices.gov/events_news/newsletters.php

Wagner, J. (1993). Ignorance in educational research: Or, how can you not know that? *Educational Researcher, 22*(5), 15–23.

Watters, A. J., Haninen, P., & Hardin, J. M. (2011). Developing a community-based research network for interdisciplinary science: The Alabama entrepreneurial research network. *Journal of Community Engagement and Scholarship, 4*(2), 67–78.

SECTION IV

Research and Writing Opportunities

CHAPTER 10

Research Agendas for Technical and Professional Communication Programs

Stephen A. Bernhardt

ABSTRACT

The value and impact of our professional writing and communication programs could be enhanced if programs set research agendas that coordinated the energies and output of their faculty. This chapter examines approaches to setting programmatic research agendas that have been successful. It identifies ways academic programs can define and engage with research activities of high value in various business and organizational settings. The chapter concludes with steps a program might take to establish a departmental or programmatic agenda for research.

INTRODUCTION

The 2011 meeting of CPTSC (Council for Programs in Technical and Scientific Communication) at James Madison University in Harrisonburg, Virginia, offered an opportunity to rethink relations between academic programs and industry. The call for proposals set a challenging agenda:

> As a professional field, technical and scientific communication overlaps both the academy and industry. Successful educational programs must therefore try to balance the interests of both areas by merging theory and practice to provide students with the knowledge and skills needed to

> succeed in a range of workplace settings (e.g., academia, education, and private and public-sector agencies).[1]

In this chapter, I suggest an approach for academic research programs to add value to the workings of academy and industry.[2] I argue that graduate programs in technical and professional communication should establish research agendas that give those programs focus and identity. Inevitably, creating a programmatic focus involves hard choices for it means refusing to be everything to everybody. Creating a research niche means hiring intentionally for critical mass vs. hiring for broad coverage or to cover what is missing because of retirements or having faculty leave. It means recruiting graduate students to a program's strengths instead of simply recruiting geographically. Importantly, creating a programmatic research niche requires a measure of central control, with some loss of freedom of choice for individual researchers, including graduate students. As I argue here, however, there are gains for those programs and for our field as a discipline.

My emphasis is on graduate programs, but a focus on specific lines of industry-relevant research would also benefit undergraduates. As Spilka (2009) notes, our professional communication undergraduates who go on to work settings need to know something of the research literature and to be familiar with methods useful to practitioners. Many kinds of research should usefully inform professional practice: usability studies, observational studies, surveys, and focus groups. Undergraduates who moved into professional communication careers could bring familiarity with lines of industry-relevant research with them, and if they were prepared to take a research-driven approach to their careers, then the hiring organizations would benefit. If graduate programs were to purposefully develop a research agenda and program identity, those actions, and the development of expertise in the department, would also influence undergraduate outcomes.

Some of my thinking derives from my consulting partner, Greg Cuppan, of McCulley-Cuppan, LLC.[3] Since 1994, Greg and I have collaborated on a range of consulting projects focused on pharmaceutical documentation in large and small drug companies. What I will term "Greg's Complaint" goes like this:

> **Greg:** Steve, what does the research say about X? I've been searching the literature and I can't find any useful studies.

[1] See http://www.cptsc.org/annual/cptsc-2011

[2] This chapter derives from comments I delivered as part of a plenary panel with Carolyn Rude (Virginia Tech) and Dan Voss (Lockheed Martin Missiles and Fire Control) titled "Building Bridges for Industry and Academe: It's a Two-Way Street." CPTSC Annual Conference, Harrisonburg, VA, October 7, 2011.

[3] See http://www.mcculley-cuppan.com

Steve: Hmmm, I don't know if that is well addressed in the literature. It would be hard to offer research-based evidence that would reliably address that issue.

Greg: Well, folks at ABC Pharma want to know the research base for our recommendations. After all, these folks are scientists and they don't want to hear our personal opinions. If they are going to change their document development processes or redesign their documentation sets, they want to do so based on research evidence. I can't stand there during meetings and training sessions telling them to do something if it is not well supported in the research literature. These folks are PhD scientists.

Steve: I'll look, but offhand, I can't think of anyone asking that question in a rigorous way.

Greg: Geez, you mean you have research programs all across the country, and in Europe, too, and nobody is asking these questions? What are you doing?

I don't think our situation is dismal, but I think we could do better by setting and pursuing research agendas at the programmatic level.

Others have voiced the same opinion, as did Blakeslee (2009) when she reported the conclusions of research forum participants asked to consider the state of research in technical communication: "The participants generally sensed, for example, that our field's research could be more coherent and cumulative and that our questions and methods could be stronger and better focused" (p. 130). This was not the first time Blakeslee made such arguments; in 2004, she and Spilka consolidated a wide range of information and experiences to assess the state of research activity in technical communication. The authors expressed the hope that "this article will inspire communities of scholars in our field to work as a collective whole toward setting goals and suggesting productive agendas for our work" (Blakeslee & Spilka, 2004, p. 75). I agree with Blakeslee that our field should be informed by a collectively defined research agenda, but I also argue that within that broad disciplinary agenda, we could profitably establish a departmental or programmatic agenda.

The most important effort to set a research agenda for technical and professional communication took place at a meeting organized at the University of Wisconsin-Milwaukee by Barbara Mirel and Rachel Spilka in June 2000, with papers and a suggested research agenda for the field published in 2002 (*Reshaping Technical Communication*). In follow-up work that relied in part on this Milwaukee meeting, Blakeslee and Spilka argue, "We need to define more and better questions in relation to industry. We found universal agreement that the needs of industry should have at least some influence on the questions we articulate" (2004, p. 78). With concerted effort, we could produce more coherent scholarship that is more cumulative and adds up to useful and usable foundations. This research agenda could benefit academic programs while providing great value to business, industry, or other work organizations. In what follows, I offer models, identify promising developments that point the way to creating high-value research programs, and offer some thinking about how to address Greg's complaint.

THE EUROPEAN EXAMPLE

A few years ago, I visited the University of Twente, in Enschede, The Netherlands, to present my work on pharmaceutical documentation and to discuss with various professional communication researchers at Twente their ongoing projects. The researchers there are familiar to many of us in the United States through Society for Technical Communication (STC) activities or other technical communication contacts: Thea van der Geest (my host), Menno DeJong, and Michael Steehouder, among others. What surprised me was the high level of central coordination of their department research program. Faculty work together to develop multiyear plans that are submitted to a national funding agency. Research funding and graduate student allotments are then conditional on the funded agenda.

The current website for the Department of Communication Science at Twente, for example, lists these research foci:

- Communication with users of technical products, services, and systems ('user support').
- The use of technical media (ICT) for communication proposes, e.g. for the communication between organisations and their public, or in health care.
- Organisational communication, in particular the communication skills of professionals in organisations.[4]

These three points represent their broad areas of focus. A plan submitted to the government funding agency would have specific and detailed studies under each broad category. Some proposals would come from individual faculty and some from departmental or interdisciplinary teams. For example, when I was there, I also met with computer science faculty interested in user experience and working with communication faculty on projects.

What's interesting is the coalescence of the agenda around user-centered and organization-centered projects. There is both a social or governmental agenda in play (healthcare communication, helping users adjust to technologies) and a business/industry agenda (organizational communication, writing in organizations). Importantly, the faculty and graduate students are aligned around

[4] See http://www.utwente.nl/gw/cw-cmc/en/

coordinated initiatives that push academic activity into corporate or government organizational settings. There is strong cross-fertilization of researchers within the department and across departments at the university, so it is very common for teams to form to pursue lines of inquiry.

The structure of research and the shared agenda mean many studies are sequential and cumulative, representing a program of research that unfolds over multiple studies and develops over time. There is an expectation that resulting research will provide demonstrable benefits to government, business, education, and community. Each of the individual research proposals is part of the larger proposal package that must pass muster before governing officials who judge projects based on social or business utility. So the proposed work of technical communication researchers is subjected to a funding agency and must appear valuable to those outside the discipline.

When I was first exposed to this agenda-setting process, I experienced some mental perturbation. What about individual faculty freedom? What about graduate students pursuing projects of their own choosing? What about non-utilitarian projects? I reflected on my own experiences in choosing exactly what I wanted to work on, with my typical funding (on a humanities model) in the form of released time. But I also realized that whenever I pursued an externally funded initiative (e.g., Illinois Humanities Council, U.S. Department of Education Demonstration Literacy Projects, New Mexico Coalition for Literacy, U.S. Department of Commerce, NSF), I was in some sense co-opted by the funding agency, and my experience and that of my graduate students were shaped by agency goals and permissible activities.

There is always some sacrifice of autonomy when working in partnership with another organization or funding agency (Bernhardt, 1992). Projects that get funded must meet the mission and objectives of the funding agency while also meeting the goals and leveraging the expertise of the applicant. So the individual must meet the organization half way. It is also easier in many ways to work solo, for there is less of the business-like activity associated with funded projects or centers of research and that claims so much time: applications, reports, personnel, budget, managerial activity, coordination with agency representatives, external evaluation. So there are inevitable trade-offs.

In research connected to my consulting work in the pharmaceutical industry, we were always heavily constrained as to what we could focus on, what information we could collect, and what we could report and to whom. To gain consulting contracts, it is typically necessary to find some common ground between the strategic goals of the business organization and one's individual interests as a professional faculty member with research questions and a career path. So there is inevitably some compromise and loss of control. It's necessary to adjust one's language, approach, and expectations to find the win/win arrangement with a business and its interests. It's also necessary to make adjustments to a corporate environment, where some information must be considered private and competitive, not open and shared, as we come to expect in academic life. Consulting in organizational environments also means working on someone else's schedule, traveling, and learning to operate in a very different culture. When academics decide to work in industry settings, then, there are many adaptations to be made and in general, a loss of the freedom that comes along with the role of solo research faculty member.

Overall, I would argue that the benefits of setting a shared agenda that allows funded activities outweigh the disadvantages. I've been able to do highly interesting work as a consultant, pursuing projects on such topics as the rhetoric of clinical study reports (Cuppan & Bernhardt, 2012), review practices in document development, and team-based technologies. From time to time, I have been able to pull in faculty colleagues on my campuses or elsewhere, and I have been able to involve graduate students in various projects. I have learned a ton about how work is done in research and development organizations, learning that feeds back into my teaching. And the work I have done has opened many academic doors, including a named professorship. I have not, however, been a part of an academic program that I could say had ever set a shared agenda. But I can easily imagine a program setting an agenda for research around such topics (e.g., document development, team technologies, and workplace genres) and finding corporate partners who would open their sites for mutually engaging research projects.

I have come to believe that there is more to be gained from setting a shared agenda than is lost in terms of freedom of choice (Bernhardt, 2002). I would argue that ethical goals of equity, of respect for individuals, or other worthy ends can be embedded in a shared agenda if faculty are thoughtful in their planning and choose to incorporate what might be viewed as "social" or "nonutilitarian" goals.

CLOSER TO HOME

Our own situations in our departments and programs, typically centered within the humanities, offer a useful point of contrast with the example above from The Netherlands. Many of our departments tend to hire based on a wide coverage model, attempting to maintain a diverse but not deep faculty bench. We typically have complete individual freedom to choose research and publishing projects, and we give that same freedom to our graduate students. Frequently, the match of dissertation topic is not particularly good between the student and the expertise of his or her director, much less other committee members. The projects that graduate students choose are often not driven by social utility, business need, or a programmatic agenda, but by the instincts, desires, and interests of those students. Too frequently, the research and publications of faculty and graduate students, all within the same program or department, do not add up to form any coherent whole that might be identified as the strength, identity, or focus of the program. This situation represents a lost opportunity.

We might contrast this scenario with the sciences or engineering, wherein labs are contingent on funding, and research programs with continuous and robust funding are able to recruit faculty to areas of strength and to fund graduate students who pursue the goals of the labs or research centers and their directors. Students make progress on their degrees by gaining research training through work on projects within a specific lab; their career goals are advanced while advancing the work of the lab. Instead of choosing their own research projects, these students align their interests with the director or other lab or center faculty to advance a group agenda. The graduate students serve a form of apprenticeship under a master—a form of legitimate peripheral participation, if you will (Lave & Wenger, 1991). They finish their degrees when they have learned enough, done enough, written enough, and can claim sufficient individual ownership of a particular study or method in an environment that is always collaborative.

One might argue that a science model would never apply to a humanities department, and especially to a big-tent English department, but I would offer some evidence to the contrary based on my experiences as Chair and faculty member of the University of Delaware Department of English. Following an external evaluation that identified the need for change, and after discussions about the unsustainability of our current configuration, we decided to work toward a more unified departmental agenda. Our old model of strength across all areas of literary studies, with coverage of both British and American periods, was too challenging to sustain. We could not hope to maintain the ranks with several faculty members in each defined period and also address other emerging and pressing concerns, such as world literatures, minority literatures, and cultural studies. There were other pressures. We were committed to expanding faculty expertise in our writing programs, including professional writing. We also had to expand support for thriving programs in English education and journalism, while determining how to support creative writing.

One response was to create a graduate program focus in print and material culture. The focus allowed us to take advantage of notable strengths at UD, including a world-class program in art conservation and preservation, while linking our faculty interests to others in art history and history. The new focus helped connect us to cultural institutions that are close by, especially the Winterthur Museum (American material culture), the Hagley Museum (American industrial culture), the Free Library of Philadelphia, and the Special Collections of our own campus library. Importantly, the move to create a focus of research and teaching allowed us to make brilliant new hires who crossed the traditional periods and nations that had tended to define older models of literary studies.

A remarkable number of faculty members began to redefine their work in terms of print and material culture, and the college found resources to stimulate cross-disciplinary research initiatives. We created summer institutes to attract more top minority or underrepresented students to the campus, efforts that are now paying diversity dividends. We worked across departments to create a summer institute for fellows in material culture studies, wherein students from various departments would hone their research and public communication skills while supported with generous stipends. We began attracting graduate students with backgrounds and intellectual interests in print and material culture studies, so we are now seeing dissertations that reflect our program's identity. Overall, we have better graduate students and a more competitive process of admissions, improved time to degree, and better placements.

What surprised me most were the ways faculty found to identify with the focus on print and material culture. We were initially worried that faculty would feel excluded if their interests were not well aligned with material culture studies. But many faculty members found ways to connect the content and materials of their classes, both undergraduate and graduate. Several faculty members, for example, realigned their research interests and found new collaborators in and outside the department. Hiring discussions, always the most divisive and contentious of meetings, improved as there was increased interest in defining positions and seeking candidates with a good fit with our programmatic emphasis.

In graduate courses across disciplines, the presence of students from English, art history, history, and art conservation and preservation helped keep those courses lively and challenging for all. New undergraduate courses were developed in book history, cotaught with our librarians. An "Introduction to Material Culture Studies" was taught by a professional writing faculty member (Debby Andrews), forming the basis for a new interdisciplinary minor, which she has directed for several years. Another professional writing faculty member, John Brockmann, connected his interests in the history of technical communication to material culture, especially his work on steamships and sailing vessels. Faculty in composition studies found their interests in multimodal composition, visual and electronic rhetoric, and online communication all benefited and were energized by attention to the print or material qualities of texts (and the nonmaterial, electronic, or evanescent qualities of texts).

I won't argue that the move to define a department research agenda solved all our issues, but it was tremendously positive and has continued to provide important benefits over the past decade. The move has elevated the national reputation of our graduate program and increased the visibility of the department in general. The initial move to identify a strategic agenda in print and material culture led to our identifying two other areas of emphasis within our graduate program. Importantly, it positioned our department as one that looked outward toward other disciplines, improving our interdisciplinary stature within the college and leading to increased resources and faculty lines.

I know this example of English at Delaware is not a success story for professional writing. My point is that a department can work together to define a shared agenda, and that agenda can connect to organizations outside itself. The move did open our department in at least some modest ways to worlds of work, as our students have become more marketable in such settings as publishing and special collections in libraries. Our students are seeing connections to collections of various sorts, and they have come to understand more about the operations of nonprofits. For those of us in professional communication, the possibilities for connecting to business, industry, and government or non-government organizations are much more obvious.

CONNECTING ACADEMIC RESEARCH AGENDAS TO WORK SETTINGS

If we scan the landscape of technical and professional communication programs, we would notice that some programs have made noticeable efforts to establish identities by setting research agendas. I would call to mind the University of Arizona and its work in community-based literacies, an area of high activity at Michigan State, where another strength is cultural rhetorics. Penn State continues to sustain a relatively small but very high quality program with deep attention to rhetorical history and theory. Minnesota has long had a cluster of high profile faculty working in the rhetoric of science. Ohio State has created a powerful concentration of faculty in digital media, as have Georgia Tech and Central Florida. Usability labs at Clemson and Texas Tech suggest a commitment to industry-relevant research, and both institutions have benefited from some studies funded by the publishing industry.

There is still room, however, for programs that want to establish particular identities tied to industry or government trends and needs. To suggest the possibilities, let me ask you, the reader, where you would send a bright student with interests in areas such as the following:

- Grant and proposal writing (or, more generally, writing for development)
- Medical writing and healthcare communication
- Document development in specific industries (e.g., aerospace, chemical, materials, drug, or energy)
- Distributed communication, meeting technologies, and distance communication, especially in connection with document development
- The challenges of building and using large and complicated texts, libraries, or textual databases
- Workplace training and development, including instructional design
- World Englishes, including globalization and localization
- Writing and communication in government, agency, or regulated settings

If you are like me, you can think of individual researchers or published reports in such areas, but it is harder to pinpoint programs that have a team of excellent faculty and a record of published research that suggests a real center of excellence. I suspect readers could add other areas to this list that might make a useful programmatic focus for research and scholarship within a department or program of professional communication.

Some years ago, the STC created a priority list of areas of needed research. The list was used to help set a research agenda for funded projects sponsored by awards from STC. Schriver summarizes this STC activity and its designated areas of research in a list that is quite useful today:

- Developing as an information designer.
- Anticipating the needs of audiences and stakeholders.
- Designing information visually.
- Writing and designing for international use.
- Teaching information design.
- Developing methods for people-centered design.
- Assessing technologies for design.
- Making decisions about media.
- Understanding trends affecting the field.
- Criticizing and refuting ideas about research. (2002, p. 113)

To compile this list, Schriver (2002) blended her own agenda, originally generated in 1989, with a second list from Brad Mehlenbacher of North Carolina State University, with the list generated by STC's research grants committee (on which I played an active role during generation of this list). STC had to suspend its program of research awards for budgetary reasons, and consequently, the list of research priorities is no longer in circulation. However, the work to set a shared research agenda in the profession was well received and undoubtedly shaped the work of various individuals. It is very suggestive for programs that would decide to set their own agenda.

We can identify research initiatives of long-term importance and impact. Many of us recall the productive collaborations of the American Institutes for Research, Document Design Center. Under the direction of Janice Redish, it developed both an outstanding body of research on plain English and document design and a focused stream of research undertaken in subsequent years by Karen Schriver and other researchers associated with Carnegie Mellon University. The concentrated effort of multiple researchers in both university settings and government centers coalesced to create a stream of important research that continues to provide value to the present day. The work of this initiative is continued by organizations like the Plain English Foundation.[5] This is work undertaken by academic researchers in partnership with a coordinating center that provides results of high value in many work settings. There are many connections

[5] See http://www.plainenglishfoundation.com/

to government organizations and a wide range of industries: telecommunications, health care, insurance, and others.

We have need of such programs of focused research. To take one example, there has never been a more concentrated focus on science communication and science literacy than now. The NSF increasingly is emphasizing that all funded projects engage in various forms of outreach to the public and to students and teachers in schools. There is a strong realization that scientists need to be able to communicate their work to audiences at all levels and that we need a scientifically literate population. Whenever graduate students are funded through NSF, such as the IGERT programs (Integrative Graduate Education and Research Traineeship), the strong insistence is on equipping graduate science students with the skills, attitudes, and experiences to be able to communicate their work broadly. A second broad opportunity exists in health care, triggered in part by the Affordable Care Act. There are research and development opportunities in electronic records systems, healthcare literacy, health insurance, and access to care.

But it is not necessary and it may be a bad strategy to rely on national agencies or government directives to determine a programmatic research agenda. The unit of decision and change is probably best located at the program level, with the faculty who determine the program's goals within particular institutional contexts. Many perceptive discussions of the relationship of the local program context to goal setting or mission setting are contained in the various *Proceedings of CPTSC*,[6] so there is plenty of existing context that can help a program begin its discussions, learn from other programs, and get a head start on self-definition. The *Proceedings* also offer many discussions on how to form productive relations with industry and what kinds of research we ought to be pursuing.

Within a department, a first step would be to take inventory not of the empty holes within a department or the need to cover certain areas (the typical starting point of a departmental hiring discussion), but of existing personnel, expertise, interests, and opportunities. In the language of SWOT analysis, the focus could first be on existing strengths and opportunities, not weaknesses or threats. The first place to look for synergistic opportunities is the people already in place, already committed to a program. Understanding existing strengths and potential areas of focus can then lead to planning for new hires and new interdisciplinary partnerships that will strengthen programmatic focus.

If a program sets an agenda, it is important to think strategically about the context of the program and institution. Faber and Johnson-Eilola (2002) provide a useful case study in their discussion of Clarkson University. Their emphasis is on creating a programmatic focus that responds to the place of knowledge workers within a global economy. In thinking through how technical communicators provide value to an organization, they note,

[6] See http://www.cptsc.org/pro.html

> If technical communicators hope to influence their (or another) company's strategic missions, they must leverage their knowledge and skills to add value to the company through new knowledge which, in turn, creates new revenue-generating products or processes. Such work would position technical communicators somewhat differently—as knowledge workers or even as consultants within their own organizations. (p. 142)

Such thinking about programs, goals, missions, and outcomes points the way for a curricular focus and a research agenda. Importantly, it ties academic work to economics and a bottom line too often absent from discussions of professional writing. Writing and communication activities are a huge uncosted activity in organizations, ripe for improvement in process and practice. Faber and Johnson-Eilola also link the academic program to broad industry trends while elevating the status of professional communicators in organizations, so there is work being done here in human potential. Theirs is just one of many possible routes a program could take.

Dray (2009) writes perceptively about the challenge of doing research that is relevant and useful in industrial settings, but argues compellingly for the value of doing so:

> For academic work in our field to play this role of supporting practice, it needs to focus on things that matter to practitioners, but it is not only the practitioners who will benefit. Such "engaged scholars" will find a wealth of extremely interesting, perplexing, and complex problems to research—great grist for those who want to make a difference in the world. Researchers who get too locked into "normal science" elaboration of existing research paradigms will miss out on new research questions of interest that will be raised by the evolution of new technologies. The following are some other areas that come to mind:
>
> - Studies of methodological validity under realistic conditions
> - Empirical assessments of common product development practices and approaches
> - A paradigm for case studies that lets us draw generalizable lessons from rich experiences in the trenches
> - Once we have this, we need to test some of these generalizations
> - Longitudinal studies of the process and dynamics of product adoption and the behavior change that goes with it, because many business decisions are based on intuitive psychological theories of how masses of people change behavior to adopt technology. (p. 6)

Drays' suggestions offer insight into what might shape a productive research agenda inside a program tied to a particular industry—she is offering her perspective on usability and user-experience research. Note that the ideas she puts on the table for productive and valued research could be applied to many industries.

A broader view is taken by Carolyn Rude (2009) when attempting to map the research areas of technical and professional communication in the form of questions. She insists the field must be able to define itself in terms of its research agenda: "Research questions, more than research methods or topics, define a field internally and externally by pointing to the knowledge making that is unique to the field. Questions are dynamic and generative" (p. 175).

In this valuable article, Rude (2009) systematically surveys our literature, drawing out the themes or emphases across many authors to define the research parameters of our field. She importantly notes that a focused research question forms the basis for an iterative program of gradually refined questions and targeted research. Rude is doing for our field what I am arguing here could be done at a programmatic level: articulating a research agenda. Interestingly, Rude acknowledges that many of the themes that form the basis of her maps were raised in department-level discussions, as faculty at Virginia Tech engaged in extended discussions in anticipation of a new PhD program in technical communication.

The important work of defining a research agenda begins at the department level. It is touchy, difficult work, for it involves reconfiguring an existing department, hiring toward core expertise, and changing business as usual. It is likely to be easier within a stand-alone writing program than within a multifunctional English department. Young, pretenure faculty probably cannot take the lead. Though they might have the most to contribute to the conversation, focusing a program will demand the leadership of the chair and other department administrators. The process can be approached incrementally, with time given to strategic discussions and coalition building. Such change is most likely to be successful if it supports college or university strategic plans and if it builds bridges of cooperation and sharing of expertise across disciplines. Deans understand the issues here and can be strong champions for change.

ACTION STEPS TOWARD A RESEARCH AGENDA

If my arguments make sense, programs might be encouraged to set a departmental or programmatic agenda for research. In closing, I will consolidate my advice here on how to go about setting an agenda:

- Start the internal conversation with department researchers. Seek common ground. Debate the pros and cons of having a focused agenda. Give plenty of time to holding these discussions under some other forum than the department hiring discussion. Consider people's feelings and issues of professional identity, but don't succumb to "we can do everything" thinking.
- Survey the contemporary research landscape in technical and professional communication. Review the literature that attempts to set an agenda for the field and that outlines issues of academic-industry partnerships. Much of that

literature is referenced here. Take advantage of the documented experience of other programs. Consider past attempts, pay attention to success stories, and work to generate options that are both defining and inclusive.

- Request your Chair or Dean coordinate a program review with external evaluators if the department has not had recently enough to be useful. As part of the process, raise issues of research productivity and focus, program identity, competition for funded research, and the cumulative impact of your department's work. Make sure your review committee is populated, at least in part, by individuals from departments that have managed to create a strong national identity.
- Evaluate the strengths of your department faculty, looking for bridging interests. Concentrate on existing resources, not what the future would hold if only you had additional hires. Do the same for other departments with cognate interests. See if you can enlarge the circle of participating faculty and potential collaborators. Interdisciplinarity is crucial.
- Talk with or survey those organizations that hire your graduates. See what you can learn about industry and organizational needs and problems. Determine ways that the mission and goals of your department might be usefully linked to contexts of work outside the academy.
- Be patient and take your time, expecting the agenda to evolve as your department evolves. Not all will go as expected, and serendipitous opportunities will emerge. Over time, your department will coalesce around an emergent agenda for research.

The future success of our field and our programs depends in part upon establishing and pursuing research agendas that are of high value to our stakeholders outside the university: business, industry, government, and nongovernmental organizations. Programs can choose to play leading roles in creating research agendas of high value. But they need to find new ways to coordinate research activities around areas of high value. This chapter has presented a framework that programs can use to foster effective academy-industry relationships that are a win/win for all involved. The author now encourages readers to consider how they might forge such relationships and move their own programs, and the overall field, in the direction of defined agendas and true partnerships to produce high-impact research.

REFERENCES

Bernhardt, S. A. (1992). *Working through conflict: Tensions in funded projects.* Proceedings of the Council for Programs in Technical and Scientific Communication. Boise, ID. Retrieved October 30, 2012, from http://www.cptsc.org/pro.html

Bernhardt, S. A. (2002). Active practice: Connecting academia and industry in technical and professional communication. In B. Mirel & R. Spilka (Eds.), *Reshaping technical communication: New directions and challenges for the 21st century* (pp. 81–90). Mahwah, NJ: Lawrence Erlbaum.

Blakeslee, A. M. (2009). The technical communication research landscape. *Journal of Business and Technical Communication, 23,* 129–173. doi:10.1177/1050651908328880

Blakeslee, A. M., & Spilka, R. (2004). The state of research in technical communication. *Technical Communication Quarterly, 13,* 73–92. Retrieved October 20, 2012, from http://www.tandfonline.com/toc/htcq20/current

Cuppan, G. C., & Bernhardt, S. A. (2012). Missed opportunities in the review and revision of clinical study reports. *Journal of Business and Technical Communication, 26,* 131–170. doi:10.1177/1050651911430624

Dray, S. M. (2009). Engaged scholars, thoughtful practitioners: The interdependence of academics and practitioners in user-centered design and usability. *Journal of Usability Studies, 5*(1), 1–7. Retrieved November 2, 2012, from http://www.upassoc.org/upa_publications/jus/

Faber, B., & Johnson-Eilola, J. (2002). Migrations: Strategic thinking about the future(s) of technical communication. In B. Mirel & R. Spilka (Eds.), *Reshaping technical communication: New directions and challenges for the 21st century* (pp. 135–148). Mahwah, NJ: Lawrence Erlbaum.

Lave, J., & Wenger, E. (1991). *Situated learning: Legitimate peripheral participation.* Cambridge, UK: Cambridge University Press.

Rude, C. D. (2009). Mapping the research questions in technical communication. *Journal of Business and Technical Communication, 23,* 174–215. doi:10.1177/1050651908329562

Schriver, K. (2002). Taking our stakeholders seriously: Re-imagining the dissemination of research in document design. In B. Mirel & R. Spilka (Eds.), *Reshaping technical communication: New directions and challenges for the 21st century* (pp. 111–133). Mahwah, NJ: Lawrence Erlbaum.

Spilka, R. (2009). Practitioner research instruction: A neglected curricular area in technical communication undergraduate programs. *Journal of Business and Technical Communication, 23,* 216–237. doi:10.1177/1050651908328882

CHAPTER 11

Theorizing a Practical Rhetoric for Virtual Collaboration among Writers in Academia and Industry

Charlotte A. Robidoux, Beth L. Hewett, and David W. Overbey

ABSTRACT

Collaborative virtual writing is a practice shared among an increasing number of individuals and organizations. This chapter analyzes the experiences of two editors and one writer who collaborated virtually on a book project involving 22 writers affiliated either with academia or industry. The collaborative process revealed that the communicative and file-sharing affordances of virtual technologies alone were insufficient to alleviate persistent conflicts between academic and industry professionals about the book's content and style. Ongoing rhetorical analyses of this collaborative process and the tensions that emerged gave rise to key principles of a practical rhetoric for virtual-collaborative writing examined in this chapter.

INTRODUCTION

Although appeals for ongoing partnerships between academic and industry persist (Blakeslee, 2001; Bosley, 1995; Ekers & Staples, 1997; Hart & Glick-Smith, 1994; Pullman & Gu, 2009; Staples & Ornatowski, 1997; Tovey, 2001; Zemliansky & St.Amant, 2007), case studies describing actual collaborations seem lacking. While some working partnerships might not be documented, an absence of examples could be due to other reasons. That is, forming such partnerships could be challenging if professionals subscribe to different affiliations (e.g., academic versus industry conferences), lack the time to establish and maintain partnerships or to read the literature, have inadequate organizational support, or retain cultural biases such as favoring theory (from an academic perspective) or practice (from an industry perspective).

As recently as an October 2011 conference at James Madison University, the Council for Programs in Technical and Scientific Communication selected the subject of academic and industry partnerships as the conference theme. The program description mentioned the challenges underlying such collaboration: "Establishing and maintaining effective relationships and partnerships between educators, academic researchers, and private- and public-sector practitioners, however, can be difficult, for different parties can have varying perspectives and understandings of what 'others' in the field do." Although these calls to action persist, examples of collaboration across academic and industry boundaries are harder to find. While interest in collaborative and workplace writing has persisted in each setting, professionals lack ample descriptions of working partnerships or models of collaboration, such as that described by Nancy Macduff and Ellen Netting (2000), between and among members of these two groups.

Because examples can offer insight into such problems and solutions, we provide an overview of a partnership that culminated in a book developed both for practitioners and academics: *Virtual Collaborative Writing in the Workplace: Computer-Mediated Communication Technologies and Processes* (Hewett & Robidoux, 2010). In this chapter, we describe this experiment in which we sought to use theory and practice to inform the book's discussion of virtual collaborative writing. Then we draw on the canons of rhetoric to explore the values that shaped the outlook and norms of the writers who contributed chapters for our book and whose affiliation with academia or industry appeared to affect writers' beliefs about and approaches to writing (see also Robidoux & Hewett, 2010). In so doing, we draw on pertinent examples from our work to present a tentative theory of collaborative writing that is practice-oriented.

OVERVIEW OF THE TOPIC

Although writing typically is considered a solitary activity, the study of its collaborative dimensions is not insignificant. Speck, Dice, Johnson, and Heaton (2008) summarized the theoretical and practical developments in collaborative writing in academic and nonacademic settings from the 1970s to 1997. A number of studies have also linked collaboration directly with the personal computer, information technology, and the Internet (Beck, 1995; Cerratto Pargman, 2003; Cerratto Pargman & Rodriguez, 2002; Selfe, 1992; Sharples et al., 1993). Other studies have identified and explored types of collaboration and why effective models for collaborative writing are essential in the workplace to improve information quality and to support decision making, processes, agreements, and overall productivity within an organization (Lowry, Nunamaker, Curtis, & Lowry, 2005; Geisler et al., 2001; Lowry, Nunamaker, Lunsford, & Ede, 1992). Still other researchers have considered the implications of increased digital technology, which have markedly changed writing paradigms (see Paradis, Dobrin, & Miller, 1985; Robidoux, 2008a; Schendel, Neal, & Hartley, 2004; Weiss, 2002; Yates, Orlikowski, & Rennecker, 1997). And in the context of social work research, Peg Hess and Edward Mullen (1995) noted that individuals trying to build collaborative partnerships must understand factors affecting decision making (e.g., relational, organizational, and political considerations that influence behavior). Within this area (i.e., social work research), Macduff, a practitioner, and Netting, an academic, (2000) described the impact of "fundamental organizational and institutional issues" on collaborative work.

Given the array of contexts and applications for collaborative writing, numerous definitions of the term are available (Hewett, Robidoux, & Remley, 2010, pp. 2–9). For the purposes of this chapter, we draw on that same definition of collaborative writing that highlights the central role of a shared productive purpose and process. Interestingly, agreeing upon how best to define collaborative writing is central for cultivating academy and industry partnerships. We thus seek to deepen the research on collaborative writing by discussing the lessons we learned and opportunities for integrating theory and practice. This discussion seeks to begin formulating a theory of virtual collaborative writing that is grounded rhetorical *praxis*.

Method of Examination

To capture the essence of our work within this context, we employed a hybrid approach, combining the rhetorical theory and aspects of case study methodology. We also drew on the views of Chaïm Perelman and Lucie Olbrechts-Tyteca (1969) to consider how a profession's system of values can shape the structure of partnerships and used the five rhetorical canons to examine our writing process. In so doing, we integrated aspects of the case study method to arrive at theoretical premises based on our experiences of working collaboratively across academy-industry boundaries. For this project, we asked contributors from both groups to employ virtual collaborative writing strategies and tools as they wrote about that subject for a dual academic and industry audience. Our aim was to improve our understanding of virtual collaborative writing by putting into practices the very theoretical concepts being considered in the book.

Throughout the project, we observed tensions on several levels—tensions related, we believe, to the cultural divide between academia and industry. In some cases, contributors became entrenched in practice-based issues that theory should have helped them to avoid. The distance between theory and practice, however, made it difficult to find solutions. Instead, the experience gave rise to a variety of questions:

- Why did our team of writers representing these two communities encounter the same "creative tension" Miller (1989) explored 20 years ago (p. 21)?
- How do competing sets of values about theory and practice make collaboration between these the two groups problematic?
- How do similar values about the importance of technology for collaborative writing ultimately lead to challenges?
- What theoretical model of virtual collaborative writing can be considered to help minimize the frictions we encountered in developing our book?

In this chapter, we examine these questions to gain insight and arrive at a practice-oriented framework for discussing virtual collaborative writing.

APPLICATION OF IDEAS

To contextualize the challenges that emerged while working collaboratively, we begin with a description of the book's inception and how we recruited an academic- and industry-based writing team. As editors, we saw our primary goal for developing a book on the subject of virtual collaborative writing as drawing on the experiences and research of the book's contributors and to learn by doing. We were committed to gathering real-time practical experience about various approaches and technologies that support virtual collaborative writing. We also believed that hands-on information gleaned from this project would enable contributors to comment authentically on the challenges that emerge when writers collaborate virtually.

Functioning as editors, Robidoux and Hewett did not presuppose that this virtual collaborative writing project would be scientific in nature regarding our findings. Rather, the project's goals were to deepen our own understanding of what it means to create an integrated, distributed writing team; to share ownership of content; and to depend on others for contributions to an overall product. Given these aims, we invited writers from the academy and industry to contribute chapters and conduct peer reviews based on their knowledge and experiences about virtual collaborative writing. Because the effort would require working together for roughly a year, we sought individuals with a deep commitment both to developments in the technical writing industry and to discovering new ideas and ways of working collaboratively.

Creating a Core Writing Team

We first reviewed proposals and accepted the work of 30 individuals interested in the project. Our intention was to promote stability and trust in this virtual environment rather than a sense of competition and fear or that someone would be removed from the project. To this end, we invited writers to comment on and revise each other's work, hoping for an ongoing cross-pollination of ideas. Yet many contributors were uncertain how to begin collaborating on such an experimental venture. Over the course of a few months, we helped those needing assistance with outlines and drafts to jump-start the process. Despite our efforts, 8 writers left the project, and of the 22 who remained with the project, we never considered dismissing any of them for inadequate submissions. We instead provided significant recommendations for reworking chapters to fit effectively with other parts of the book.

With a smaller group, we promoted a collaborative decision-making process around the selection of tools. The effort we expended on tools so early in the project underscored not only the challenge of finding effective processes and technologies for a team of writers collaborating virtually, but also the fundamental need for trust among team members. Without trust, our team would struggle to reach agreement about what tools to use. As a result, the struggle itself functioned as an important catalyst for team members to become acquainted, develop working relationships, and establish trust. Eventually, we decided on a collaborative wiki for initial topic and book development, and our tool selection was intended to support both synchronous and asynchronous writing practices that allowed writers to track changes and respond to feedback.

We gave writers 3 months to complete their first drafts, inviting them to brainstorm and share ideas on a wiki and a dedicated listserv. As the project progressed, we had to prompt writers repeatedly to provide feedback to each other using one or more of these tools. Our goal was to place less importance on individual content development and ownership and greater emphasis on the project's overall content integration. However, this collaborative approach was so atypical for many writers that reluctance to share ideas was pervasive. As book editors, we realized that overcoming this introversion would require a continual effort to develop effective collaborative habits.

Questions of Value in Acts of Collaboration

Having worked through the decision about tools, writers needed to position their chapters within the structure of the book and grow their ideas collaboratively. Certain challenges, however, soon emerged. These challenges often involved academics and industry writers divided by profession and polarized on various positions. As editors, we soon realized this divide was related to each group's adherence to a set of values—what Perelman and Olbrechts-Tyteca (1969) called potential "objects of agreement" (or disagreement) (p. 74).

The underlying attitudes, operating assumptions, and norms that separated the groups influenced their perceptions of and approaches to collaborative writing. An e-mail written by an industry professional highlights these opposing values around theory and practice:

> I don't have much in the way of "theory" to refer to as the basis for what I am going to talk about. What I have is "practice" and a set of principles drawn from practice that can help guide the development of similar practices in

> other fields. It is not unreasonable to call such principles "theories." On the other hand, this may not be what academic readers expect when they say the word "theory," and "theory" is probably not the first word that springs to mind for an industrial practitioner either.
>
> Industry develops industrial practices. It judges those practices based on results, retaining those that work and discarding those that fail. To be certain, academic studies of these practices may result in the development of theories that explain why they work, but such studies are by no means always available, and even if they are, they do not necessarily flow back to the industrial community that developed the practices. In many cases, the study and dissemination of these practices are not done by academics but by consultants and the books they write are not academic treatises but business books. . . . I think that in the industrial world "principles and practices" play a role similar to the role played by "theory" in the academic world.

This passage reflects strongly held opinions about a "structure of reality" shaped by industry-based experiences (Perelman & Olbrechts-Tyteca, 1969, pp. 261–263). This e-mail compares concepts of theory and practice as they relate to both environments, looking for intersections between seemingly opposing intellectual processes. The need to moderate the entrenched structures of reality about theory or practice persisted throughout the development of the book.

Within our team of writers, academic writers clearly relied on theoretical constructs as a basis for understanding virtual collaboration, whereas industry writers trusted their intuition and practical applications. In both groups, organizational and philosophical influences reinforced the academic bias toward theory and the industry bias toward practice, corroborating Hess and Mullen (1995) and Macduff and Netting (2000). While we did not know if academic institutions would value or support a collaborative research project, the many participants from academia indicated that such projects were valuable. On the industry side, we hoped that organizations would allow practitioners to participate in collaborative research. In most cases, industry writers were free to participate as long as they worked on their own time and preserved organizational trade secrets.

Examining Our Writing Process Through the Canons of Rhetoric

As these cultural differences emerged and intensified over the course of 4 months, we realized how much we needed to draw on our own rhetorical skills to reframe ideas and strategies as compelling objects of agreement that would help to advance our efforts. What follows is an overview of the discernible, cultural divide related to theory and practice within our team. To understand the value differences that shaped the book's development, we drew from the five canons of rhetoric.

Invention

Invention became a discussion topic when considering ways of preparing or guiding writers through a collaborative process. For teachers of writing, invention entails helping students to find an argument, to discover what to say, or to amplify a topic. Managers in industry, however, seek concision over idea fluency because tomes of text take more time, cost more money to translate, and confuse users who need quick access to answers.

For this project, contributors had to explore a narrower view of invention, one with a different kind of creativity. Rather than finding creative ways to argue points, technical writers who write reusable topics focus on developing them according to particular standards (so-called information models) and limiting phrases to ones already written or translated. This point is important because product distribution is increasingly global and requires translation (or localization) into many languages. To make information easier to find and more cost effective to translate, companies train writers in the practice of "text reduction," "minimalism," and "standardization" (Hackos, 1999, p. 17). But some technical writers believe this trend limits creativity, which challenges documentation mangers to help writers redefine creativity.

When analyzing their audience, technical writers may unknowingly use common and special topics to develop concise units of information such as concepts, procedures, examples, or illustrations (Miller, 1987, p. 67; Miller & Selzer, 1985, p. 311; Robidoux, 2008a, p. 113). For writers in industry, however, it can be difficult to convince engineers, subject-matter experts, or those in positions of authority to develop ideas in a certain way to make content more useful to a customer, as one writer on our project suggested,

> Initially I was reluctant to even comment on the wiki in the presence of others—especially the academicians—who, I assumed, must know more than I do. I got over this, but it was reminiscent of the reluctance we see in my department by some writers who are afraid to speak up in front of more senior writers, or others whom they perceive as "the experts" in some domain.

Thus, it seems logical that technical communicators must balance their own expertise with the authority of subject-matter experts.

Moving beyond the role of a scribe is important to ensure readability (i.e., the right amount of information and level of detail) for a given audience. Accordingly, writers must rely extensively on a range of heuristic strategies to devise cohesive document structures, concise topics, and convincing arguments about how best to deliver information. A better understanding of invention enabled our contributors to collaborate across boundaries. While academy writers learned to scale back the development of theory, industry writers learned to expand their rationale for arriving at certain conclusions and recommendations.

Arrangement

In the process of planning our book, we envisioned a typical "academic" organization of ideas, one that began with theoretical underpinnings. This strategy biased the scales of collaboration toward the academy even though the book was planned for a dual audience. Although both editors have industry experience, their training as scholars somehow took precedence. An industry contributor, who eventually left the project because the cultural differences between academia and industry seemed too problematic, explained:

> I've been pondering the structure of the first section of the book [Background and Theoretical Grounding], and its relationship to the second section [CMC Technology and the Rhetoric of Collaborative Virtual Writing]. They are set up more or less as theory followed by practice, which is fine from an academic point of view, but it is seldom how things work in industry. . . . Since the aim is to appeal equally to both audiences, I wonder if we should broaden the scope (or at least the title) of the first section to include "principles and practices."

Because the editors originally encouraged a typical academic book structure (i.e., theory followed by discussion and illustrations), they inadvertently triggered the rift between theory and practice. Academic writers appeared to feel more comfortable jumping into a discussion about the literature on the subject and synthesizing summaries of seemingly relevant theories of collaboration. However, industry writers (outnumbered by the academic writers 2 to 1) expressed that they had little to offer to the discussion, and some indicated that they felt intimidated or alienated by theory that seemed far removed from their experiences.

An industry writer, who finally joined the discussion on a listserv e-mail, announced her nonacademic affiliation up front and shared her opinion about how to integrate theory and practice to make progress on the book:

> First, please be aware that I'm not an academic, but I thought I would weigh in from a practitioner's standpoint. . . . I guess we are saying that some (?) degree of a priori knowledge is needed, with a healthy dose of a posteriori principles? IMHO, the sooner we can summarize the former, the greater the likelihood we can provide supporting examples. Conversely, from the case studies, we can induce the principles. I guess it's a question of maintaining an appropriate balance.

This practitioner appealed to academic values through word choice, logic, deference, and compromise. In sum, while scholars were taking their time tossing around theoretical constructs, industry writers wanted to get to business. Such a difference could be based on the cultural norms for writing in one context versus another. Scholars, who get paid to explore ideas, have more decision-making power or influence over how their time is allocated or when deadlines occur. In contrast, companies pay practitioners to deliver prescribed amounts of content at regular intervals in support of product delivery schedules. Often, industry writers have little control over their schedules.

The concern about overloading the book with ideas resulted in an online discussion on our wiki. The point and counterpoint considerations created a theme that the editors drew upon to present a hybrid approach that would

- Scale back the theory to ensure that the book would capture the interest of both business and academic audiences
- Combine reviews of literature and theory into one chapter, renamed, "Frameworks for Talking about Virtual Collaborative Writing"
- Limit the amount of theory in chapters by pointing readers back to the first part of the book for theoretical information
- Rearrange the planned case studies, interspersing them throughout the whole of the book instead of toward the end of the book

Collaborators thus observed the accommodation of opposing sets of values. These efforts helped build trust and enabled a greater willingness to collaborate.

Style

A conflict of values also was apparent in our writing team's uses of language (e.g., word choice, syntax, and terminology). While linguistic variation could easily characterize the speech across professions, it was surprising to see variability among individuals within the same group who shared the same values. Industry writers, for example, struggled over terms related to managing content (e.g., "copy and paste," "repurpose," "tweak, and "reuse"). After a number of e-mail exchanges, one writer drew on industry *ethos* to make her point:

> I just found JoAnn Hackos's definitions . . . the ones we use at [my company]:
>
> - Reuse refers to the process of using chunks of content in more than one context or document.
> - Repurposing refers to delivery [of] the same information in more than one medium: print, HTML, help, and so on.
>
> re: "copy/paste and tweak" I don't have a term for it, other than copy-and-paste. But I do feel strongly that we ought to make some sort of statement that this does not constitute reuse. In fact, it's more or less the opposite.

Another writer's reaction was to stick with her own use of terms or to develop a standard for all to follow, which created angst about defining terms:

> I propose we use the definitions in my chapter as they currently stand. However, if you wish to use other terms/definitions, please let me know so

> that I can make the appropriate modifications. I will need a term for "the act of copying content from one context and pasting it into a second context, optionally customizing the content for suitability in the second content, and resulting in two separate instances of the content." If you don't like repurposing, could you please provide another suggestion?

This question then prompted the first writer to contact the editors for assistance:

> I don't have a good read on this, writing to you privately for your guidance. (I'm thinking of all those trust issues with attempting to collaborate by e-mail with people you've never spoken with.) I have the sense that . . . I've overstepped the bounds by asking her [the second writer] to change her definitions to the Hackos ones, which I thought I'd introduced as more widely used. Am I misreading? Am I wrong?

Within this context of divergent expectations and definitions, the phone became essential to managing differences. During calls with authors, the editors planned to reiterate statements that would help writers realize their ideas were actually in agreement.

Memory

In the context of virtual collaborative writing, *memory* concerns how information is recorded electronically, and it can pertain to storage devices, databases, archives, software for managing style and terminology, metadata strategies, and taxonomy schemes. Our team was given a unique opportunity to manage terminology electronically because an industry writer offered free access to a quality management tool: acrolinx. This tool would help consolidate decisions about terminology and usage rules in a database we could access online. As chapters became complete, writers could run the acrolinx software to check things like terminology (e.g., terms classified as "deprecated" could be replaced by "preferred" terms). Thus, this tool helped us to ensure a coherent voice across the book.

To orient our users to the tool, we set up two demonstrations; after the first session, some academic contributors expressed concerns that using this kind of tool to fix and prescribe word choices would disrupt the creative process and interfere with the stylistic variations:

> I assume that this software [acrolinx] would be used at the later stages of the writing process. Also, as a general note about the demonstration: my question about different writing styles and the application of the rules to the different styles and voices remains unanswered, but I will do my best to work within the collaborative parameters.

Ultimately, the offer to "work within the collaborative parameters" became moot since the writer of this excerpted communication eventually dropped off the team. His choice of words illustrates the overlap that exists between the canons of style and memory in automated technical writing environments. It should be noted that industry writers also engaged in the conversation, explaining that the software was intended simply to strengthen our collaboration. Promoting the tool, however, did not influence the opinions of those who viewed it as an impediment. The divided reaction, along with the sense we were overloading the team with tools, prompted the decision that the editors alone would use acrolinx to manage terminology and style—effectively standardizing language more by editorial fiat than through cooperation.

Delivery

The knowledge that this book was intended for academic and business audiences surfaced frequently. We all raised the question, "How can we deliver a message about virtual collaborative writing most effectively?" On the business side, one writer promoted the use of examples as a more successful way to deliver ideas about collaborative writing:

> I'm a great believer in the power of storytelling, and if you look at the most successful business books (Gladwell [2008] Tipping Point, Weinberger [2007] Everything is Miscellaneous, etc.) you will see that every chapter is built around a story. Stories provide a concrete foundation for the rest of the discussion, and keeping the stories close to each part of the principles and practices discussion helps keep things in focus.

That a practitioner not formally trained in rhetoric offered this recommendation is interesting because it might be easy (and clearly reductive) to relegate the storytelling metaphor to academics alone.

Notably, our team's rhetoric scholars (the editors included) had to be reminded of the dual audience. This reminder prompted our team to work collaboratively to determine how to deliver ideas that would appeal to a business audience. The editors scheduled two conference calls and set up a virtual meeting space wherein writers could answer questions, debate options, and suggest ideas using a whiteboard format. From these meetings, the writers agreed to a new strategy for presenting ideas: incorporating opening illustrations in each chapter. Delivering information in this way would enrich the rhetorical appeals by drawing on the expertise, or *ethos*, of *all* writers as well as their experiences, or *pathos*, to create opportunities for adherence.

During these meetings, some writers tried to articulate the differing sets of values and motivations that came into play while creating this text. One contributor arrived at this assessment:

> One thing this project has brought home to me is that there are different motivators in industry and academe, and that affects how we collaborate, and what we value in it. In industry—especially in today's agile projects—deadlines and efficiencies have at least equal weight with achieving quality. Not that we don't value and strive for quality—of course we do—but in our collaborations, often the primary drivers are efficiency and the ability to meet deadlines, while still achieving "good enough" quality.
>
> My impression as an industry writer is that in academia, while there are certainly deadlines, the driving factor for collaboration is some improvement in quality—a broader set of ideas for researchers to meld, greater learning opportunities for students, etc. Of course we have that in industry too—and each role in a collaboration has responsibility for some aspect of quality. . . . But all of that operates within an environment where the level of quality in the content you deliver is tempered by the requirement to meet release dates and control cost.

What became clear about our project is that the idea of virtual collaborative writing was compelling enough to draw over 20 volunteer writers to engage in a writing project that lasted almost a year. Among other things, this work means recognizing that in the digital era there are shared values between academia and industry, not just opposing ones. The challenge entails examining the rhetorical context, identifying perceived incompatibles, and finding ways to remove them.

THE NEED FOR A PRACTICAL RHETORIC IN VIRTUAL COLLABORATIVE WRITING

As illustrated, the experience of the editors and writers collaborating virtually on a book about virtual collaborative writing reveals a classic cultural divide between academia and industry. Whereas participants from academia emphasized theory and extended intellectual deliberation when collaborating virtually, industry participants focused on the need to generate practical solutions quickly. Further reflection, however, suggests a more complex interpretation about values that linked the two distinct groups. Both academics and practitioners—and certainly the editors—appear to have overestimated the capacity of digital media to transcend the frailty that embodies human communication. How people write and collaborate do not change drastically just because of the availability of virtual collaborative technologies. In fact, the attention to the technology itself can exacerbate rather than alleviate the difficulties writers face in a collaborative project.

Two Groups Separated by a Common Language

If this project suggests that the cultural divide between academia and industry remains where it was over 20 years ago, per Miller (1989), then we have to be cautious and recognize that tensions might become even more challenging with the addition of the unknown qualities of virtual technology. Our reflection on this project suggests that both academia and industry share values that can overstate the importance of technology (i.e., a "silver bullet" of the information age). Indeed, technology as an agent of progress and productivity is so potent a symbol because it seemingly empowers people with untold access to information and the potential to solve problems (Weaver, 1985).

Both the shared and opposed values that emerged in our project indicated that these issues concerning virtual collaborative writing were rhetorical, not technological. That the collaboration turned virtual through the use of computer-mediated technologies did not override the influence of values that the collaborators brought to the project, *regardless of whether those values were shared or in opposition to one another*. In fact, both the cultural divide between the two groups and their shared values leading them to overestimate virtual collaborative technology's capacity to overcome human limitations served to diminish their level of collaboration.

In truth, the needs of industry and the backgrounds of academics led to incompatibilities between the two groups that could not be overcome by the technology mediating the collaboration. Rather, for the project to succeed, the book contributors (the editors especially) needed to identify points of negotiation and to create hierarchies of values as a way to address competing priorities (Perelman & Olbrechts-Tyteca, 1969, p. 80). The editors conducted many individual or small-group phone conferences (using seemingly unsophisticated, old technology) to discuss issues that were frustrating team members and to find options for compromise. In other words, the editors had to address a rhetorical problem with rhetorical solutions that demoted the importance of virtual technologies.

These observations reveal the educational richness in this project between academia and industry, for they complicate current assumptions about the capacity of virtual collaborative technology to bring together these two groups and in turn establish possibilities for what the two groups have to offer one another. The questions then become (1) How do academics and practitioners collaborate with one another? and (2) How can they engage in practice that generates knowledge that is neither exclusively academic nor nonacademic?

Miller (1989) seemed to address this factor when writing that

> understanding practical rhetoric as a matter of conduct rather than of production, as a matter of arguing in a prudent way toward the good of the community rather than of constructing texts, should provide some new perspectives for teachers of technical writing and developers of courses and programs in technical communication. (p. 23)

In essence, it's a matter of community.

If the *community* in question excludes practitioners, once again, an inclusive perspective morphs into an exclusive one. The concern for "the good of the community" seems relevant for meeting academic needs—the needs of teachers and course developers. But "the good of the community" must extend beyond the needs of academia. Miller's cited reasoning exposes the Achilles' heel of academia: Its contribution to the "good of the community" is determined by the academy itself rather than the community-at-large.

The mistake academia often makes is to conflate "knowing about nonacademic practices" with being just a "receptacle for practices and knowledge created elsewhere" (p. 23). To enable the community good through the teaching of writing, the academy should be theorizing about how and why nonacademic practices succeed or fail. Such investigation would leave behind pedagogy that treats writing as something concerned only with inquiry and criticism based on academic practices. Otherwise, academia risks being left behind if it fails to acknowledge writing as something increasingly collaborative in the digital age. (Such a perspective is arguably unorthodox for an academic culture that still values writing overwhelmingly as an individual pursuit directed toward obtaining a degree and tenure.)

Interestingly, when it comes to practical rhetoric as it relates to this virtual collaborative writing project, academia and industry appear to suffer from sharing the same faulty value system: overconfidence in the technology and oversimplified perceptions of human communication. What drives academic and nonacademic practices are not just opposing values, but shared ones as well. Both groups value the practice of writing as it functions across various rhetorical contexts. In addition, the culture in which they operate influences both groups in ways that are sometimes similar, for instance, in the way that both groups spar over issues related to style and terminology. This shared view makes academic and nonacademic communities collaborative entities by default. Our project on virtual collaborative writing shows that while academics and practitioners are in fact not identical in their needs and backgrounds, they draw on overlapping assumptions.

Nonacademic practice does not happen "elsewhere," especially at a historical moment when much writing, regardless of its relation to academia or industry, takes place virtually in digital environments. Learning about the limitations of these media and writers' experiences with them regardless of background is certainly in the best interest of both academia and industry. Regardless of the extent to which the editors' goals and expectations were met, this project signifies the need for a conscious and explicit collaboration between academia and industry. Specifically, both groups should examine the basic assumptions underlying writing practices in both settings. In the context of these assumptions, the groups should investigate how virtual collaborative technologies and processes will affect performance and their interactions as collaborators.

What we are dealing with in the 21st century are professional writers encountering all kinds of unforeseen difficulties and possibilities that affect everyone involved. Nonacademic practice is a practice that happens *everywhere*. Virtual environments, unlike predigital ones, are not confined to the physical, material settings of various workplaces, including the university. Learning and living the art of rhetoric, regardless of the workplace setting, is essential for collaborative writing, especially to the extent that technological sophistication makes digital interaction so commonplace. The digital age reveals that adherence of minds *is* collaboration.

Moving Beyond Epideictic Perspectives of Technology

From this virtual collaborative writing project, we now believe that both academia and industry must reexamine what we see as epideictic perceptions of computer-mediated processes and technologies. To the degree that epideictic rhetoric gives praise to so-called amazing technological capabilities of digital media, neither group currently can claim to have autonomy or to transcend the depictions of this technology in popular culture. The influence of such depictions is important; the goals and expectations of writers in both academia and industry are determined to some extent by the assumptions about the capacity of virtual collaborative technology to mediate successful collaboration. Instead, we suggest a need for a practical rhetoric that would be forensic and deliberative in nature; indeed, both academia and industry have familiarity using these discourses.

We suggest that such a practical rhetoric be deliberative in terms of future actions: How should digital media be used in virtual collaborative writing? What future courses of action should collaborators take when it comes to virtual collaborative writing projects? Those committed to collaboration should move beyond the popular epideictic discourse, which can presuppose that digital media will address or overcome the values that distinguish groups culturally. Rather, value differences can intensify when writing collaboratively in virtual settings. Effective collaboration thus requires finding ways to penetrate and mediate the value differences that make up the rhetorical contexts of the individuals collaborating.

Without a forensic and deliberative rhetoric to work through value differences, future virtual collaborations are likely to encounter the same problems that arose in this project. The same consequences are inevitable for any such project that approaches virtual collaboration from a technological rather than rhetorical perspective. As Robidoux (2008b) explains, part of the reason that values are rhetorical is that they determine the approach groups take toward a project. Additionally, questions raised about approaches to take are deliberative, requiring forensic data as a basis for negotiating to arrive at a resolution. Miller's (1989) "standpoint for inquiry and criticism" (p. 23) with regard to this challenge is neither academic nor nonacademic but inherently collaborative, because the challenge is equally important to the professional success of both groups, and meeting it requires attention to the successes and failures that writers from both groups encounter. For either group to ignore the others' experiences with virtual collaborative writing would delay and undermine the articulation of a collaborative practical rhetoric. Both groups need such a practice to put virtual collaborative technologies to best use consistently—a goal both academia and industry share regardless of differences in their professional protocols or cultures.

In rhetorical terms, the shared values among project participants reflect Jeanne Fahnestock's (1998) observations of technical and scientific discourse being "accommodated" in public literature (p. 333). That is, the process of conveying scientific findings in lay publications (because of a change in genre, audience, and purpose) typically results in an epideictic adaptation of the original findings. Contributors approached collaboration with an unexamined epideictic perspective of virtual collaborative technology, that is, an accommodated one. This perspective seems to have been shared equally by both academics and practitioners. In addition to being unexamined, such an epideictic perspective also would seem to be a departure from the forensic or deliberative perspectives that typically characterize the work of both groups. In some respects, both groups lost touch with their professional *ethos* (i.e., credibility) in their approach to this project. They operated, at times, more like anonymous members of the public rather than trained professionals distinguished by a particular expertise.

What might seem like surprising behavior might be explained if we consider work habits and a shared area of interest. That is, because both groups had experience writing books, they assumed that the familiar work style they brought into the project was common. Additionally, because both groups shared the same research interest and faith in technology, they were unprepared when confronted with practices that seemed foreign. Indeed, that both groups were focused on the same subject distracted them from the idea that their values related to this subject area could be so different.

Fahnestock's considerations also seem highly relevant to our project because she addressed, and in some ways anticipated, the need for "various segments of our society to keep up with one another" in a world where "print can seem too slow" and information sharing, even in the mid-80s, was being done "on line" (1998, p. 347). In a word, she was writing about *collaboration*, and we think it is somewhat ironic that the accommodation of discourse about which she wrote many years ago applies to perceptions today about virtual collaborative technologies.

Moreover, the epideictic celebration of these technologies by both academia and industry has recast the disconnect Fahnestock described between forensic and deliberative discourses on the one hand and epideictic discourse on the other. That is, the "various segments" at work in our project, representatives of academia and industry, tended to lapse into epideictic notions of virtual collaborative technologies. Our team did not question the assumptions we made about technological advancements that seemingly would enable breakthroughs that lead to indisputable improvements in human life. Our team's assumptions steered us away collectively from the forensic and deliberative perspectives that would sustain a constructive, knowledge-building approach to virtual collaborative writing. We also see a close parallel between the "practical system" of which Fahnestock (1998, p. 345) wrote and the "practical rhetoric" that Miller (1989) identified as essential to academic and industry approaches to writing. Ultimately, the lack of both a practical rhetoric and a practical system of virtual collaborative writing explains why intelligent professional writers reverted at times to novice behavior over the course of the project.

BEST PRACTICES FOR VIRTUAL COLLABORATIVE WRITING

When setting up partnerships between academia and industry, what are the best practices that will support virtual collaborative writing involving members of academia and industry? Six principles emerged from working on our book, based on various challenges, observations, and realizations. Because the context and circumstances for collaborating often vary, teams can use these principles to promote best practices in various workplace environments.

Principle 1—Develop a Culture of Collaboration

The collaborative mission must be clear to all members of the team. Because writers often are accustomed to working independently, expectations about the collaborative norms and practices need to be explicit and illustrated repeatedly. The cultural shift takes time and requires a commitment to managing change. Examine the norms, expectations, and goals that define the values of each group up front so that participants can consider them self-consciously as the work unfolds and as challenges arise.

Principle 2—Find and Promote Leadership

Strong leadership is essential for creating a collaborative culture. Leaders are crucial because they are ambassadors of change for the team. Smaller teams should have a single leader, but larger teams may need to divide into smaller ones with leaders identified for each. To collaborate effectively, leaders need to endorse explicit processes, roles, workflow strategies, and if possible limit the size of the team to fewer than five.

Principle 3—Establish Trust

Without trust, teams will be unable to collaborate effectively. Establishing trust takes time, and the amount of time can vary from one person to the next. All individuals need to believe that authenticity, support, and transparency permeate throughout the team. Establish guidelines for communicating and interacting. Help both groups gain insight into the aims of each other's work.

Principle 4—Use Tools and Collaborative Modes Effectively

Knowing what tool to use when can be confusing, and using the wrong one in a situation can intensify the confusion. Consistent processes about how to use various types of tools is essential to ensure stability on the team. Determine what familiarity each team member has with technology and build processes around a common set of tools. Simplify the use of technology whenever possible.

Principle 5—Create Structure

Explicit roles and carefully structured processes will create consistency and trust so that teams can maximize their use of technology. Defining and testing processes with input from team members will ensure that tools do not interfere with the team's productivity. Find processes that support various types of work habits.

Principle 6—Measure and Track Performance

Leaders will ensure productivity by measuring collaborative performance throughout the course of a project. With data from the team, leaders can make course corrections to keep the team on track. Surveys, interviews, and the quality of work products all enable teams to measure and track their efficiency. Find ways to gather information about team performance initially and during in the project to make course corrections as needed.

CONCLUSION

The friction between academia and industry that fueled the cultural clash between our two book project groups shows that their interactions are perhaps more important now than ever because of the digital era in which they both operate. The strain that editors and writers experienced raises questions about the values that drive work practices, the uses of digital media, and what it means to collaborate in the digital age. The persistence of these tensions suggests that continued collaboration among academia and industry is necessary for successful virtual collaborative writing in the future. If the digital era signals an entry into a globalized network of communication, then the future of collaboration would appear to involve interactions among an unprecedented number, variety, and combinations of collaborators than ever before.

Insight into the kinds of interactions that are possible in the future will come from continued conversations between both groups. We see professional organizations that have an ongoing interest in expanding industry-academic partnerships. For example, the Center for Information Development Management (CIDM), which was created to address the concerns of practitioners developing information globally, has encouraged a research agenda that is focused on practice.

In 2012, the CIDM Best Practices Conference assembled a panel, "The Researchers, Educators, and Communicators Roundtable: Forming a New Habit of Collaboration," to enable a reciprocal relationship between researchers and practitioners: "In a reciprocal relationship, research questions are identified in practice and then researchers design studies to address those questions. Study results are then fed back into practice to improve practice, at which point new research questions are formulated" (Andersen, 2013a). The panel's aim was to find new ways to sustain this reciprocity by creating awareness through articles, by gathering data through surveys to determine what research is needed to help practitioners, and by setting up a special LinkedIn subgroup to the CIDM group called "Academic/Industry Collaboration," which has over 50 members from both groups.

These efforts have established a springboard for research ideas, including "the need for large quantitative and qualitative studies that examine specific practices and behaviors across a wide range of users, organizations, and industries" (Andersen, 2013b). Some research subjects might pertain to user behavior, process/practice, content strategy, metrics/measurement, technical communicator roles, and training. The CIDM organization and other groups, such as Council for Programs in Technical and Scientific Communication and the Society for Technical Communication Academic Special Interest Group, have provided opportunities for sustained conversations among researchers, educators, and practitioners.

The research discussed in this chapter reflects our commitment to continued discussion around best practices for virtual collaborative writing. The reflections about our project have indicated the need for a practical rhetoric of virtual collaborative writings that is forensic and deliberative, one that moves beyond the current epideictic-like, celebratory one. To this end, one of the great successes of this project is that it introduces the possibility that in the digital age, collaboration has become an increasingly complex and rich rhetorical practice. Indeed, rather than a technologically simplified or superficial practice, collaboration in the 21st century has the potential to integrate a widening range of overlapping perspectives and formerly distinct structures of reality. This rhetorical practice requires a resilient yet determined adherence of minds willing to navigate the virtual boundaries that largely still separate academia and industry.

REFERENCES

Andersen, R. (2013a). Results of the 2012 Best Practices Research Needs survey. *CIDM Information Management News April 2013*. Retrieved from: http://www.infomanagementcenter.com/enewsletter/2013/201304/second.htm

Andersen, R. (2013b). The value of a reciprocal relationship between research and practice. *CIDM Information Management News May 2013*. Retrieved from: http://www.infomanagementcenter.com/enewsletter/2013/201305/second.htm

Beck, E. E. (1995). Changing documents/documenting changes: Using computers for collaborative writing over distance. In S. L. Star (Ed.), *The cultures of computing* (pp. 53–68). Oxford, UK: Blackwell.

Blakeslee, A. M. (2001). Bridging the workplace and the academy: Teaching professional genres through classroom-workplace collaborations. *Technical Communication, 10*(2), 169–192.

Bosley, D. (1995). Collaborative partnerships: Academia and industry working together. *Technical Communication, 42*(4), 611–619.

Cerratto Pargman, T. (2003). Collaborating with writing tools: An instrumental perspective on the problem of computer-supported collaborative activities. *Interacting with Computers: The Interdisciplinary Journal of Human-Computer Interaction, 15*(6), 737–757.

Cerratto Pargman, T., & Rodriguez, H. (2002). Studies of computer supported collaborative writing: Implications for system design. In M. Blay-Fornarino, A. Pinna-Dery, K. Schmidt, & P. Zaraté (Eds.), *Cooperative systems design: A challenge of the mobility age* (pp. 139–154). Amsterdam, The Netherlands: IOS.

Council for Programs in Technical and Scientific Communication (CPTSC). (2001, October 6–8). *Academy-industry relationships and partnerships*. 39th annual meeting of the Council for Programs in Technical and Scientific Communication. Retrieved October 20, 2012, from http://www.cptsc.org/annual/cptsc-2011/index.html/; http://www.cptsc.org/

Ekers, P. S., & Staples, K. (1997). Collaborative conflict and the future: Academic-industrial alliances and adaptations. In S. Selber (Ed.), *Computers and technical communication: Pedagogical and programmatic perspectives* (pp. 375–388). Greenwich, CT: Ablex.

Fahnestock, J. (1998). Accommodating science: The rhetorical life of scientific facts. *Written Communication, 15*(3), 330–350. Reprint of 1986 original from same journal.

Geisler, C., Bazerman, C., Doheny-Farina, S., Gurak, L., Haas,C., Johndan, J.-E., et al. (2001). IText: Future directions for research on the relationship between information technology and writing. *Journal of Business and Technical Communication, 15*(3), 269–308.

Gladwell, M. (2008). *Tipping point*. Boston, MA: Little Brown.

Hackos, J. T. (1999). An application of the principles of minimalism to the design of human-computer interfaces. *Common Ground, 9*, 17–22.

Hackos, J. T. (2007). *Information development: Managing your documentation projects, portfolio, and people*. New York, NY: John Wiley & Sons.

Hart, H., & Glick-Smith, J. (1994). Training in technical communication: Ideas for a partnership between the academy and the workplace. *Technical Communication, 41*(3), 399–405.

Hess, P. M., & Mullen, E. J. (Eds.). (1995). *Practitioner-researcher partnerships: Building knowledge from, in, and for practice*. Washington, DC: National Association of Social Workers.

Hewett, B. L., & Robidoux, C. (Eds.). (2010). *Virtual collaborative writing in the workplace: Computer-mediated communication technologies and processes*. Hershey, PA: IGI Global.

Hewett, B. L., Robidoux, C., & Remley, D. (2010). Principles for exploring virtual collaborative writing. In B. L. Hewett & C. Robidoux (Eds.), *Virtual collaborative writing in the workplace: Computer-mediated communication technologies and processes*. Hershey, PA: IGI Global.

Lowry, P. B., Nunamaker, J. F., Curtis, A., & Lowry, M. R. (2005). The impact of process structure on novice, virtual collaborative writing teams. *IEEE Transactions on Professional Communication, 48*, 341–364.

Macduff, N., & Netting, F. E. (2000). Lessons learned from a practitioner-academician collaboration. *Nonprofit and Voluntary Sector Quarterly, 29*(1), 46–60.

Miller, C. R. (1987, Winter). Aristotle's "special topics" in rhetorical practice and pedagogy. *Rhetoric Society Quarterly, 17,* 61–70.

Miller, C. R. (1989). What's practical about technical writing? In B. E. Fearing & W. K. Sparrow (Eds.), *Technical writing: Theory and practice* (pp. 14–24). New York, NY: Modern Language Association.

Miller, C. R., & Selzer, J. (1985). Special topics of argument in engineering reports. In L. Odell & D. Goswami (Eds.), *Writing in nonacademic settings* (pp. 309–341). New York, NY: Guilford.

Paradis, J., Dobrin, D., & Miller, R. (1985). Writing at Exxon ITD: Notes on the writing environment of an R&D organization. In L. Odell & D. Goswami (Eds.), *Writing in nonacademic settings* (pp. 281–307). New York, NY: Guilford.

Perelman, C., & Olbrechts-Tyteca, L. (1969). *The new rhetoric: A treatise on argumentation*. Notre Dame, IN: University of Notre Dame Press.

Pullman, G., & Gu, B. (2009) *Content management: Bridging the gap between theory and practice*. Amityville, NY: Baywood.

Robidoux, C. (2008a). Rhetorically structured content: Developing a collaborative single-sourcing curriculum. *Technical Communication Quarterly, 17*(1), 110–135.

Robidoux, C. (2008b). *Human genome project discoveries: Dialectics and rhetoric in the science of genetics*. Retrieved from ProQuest Dissertations and Theses database. (AAT 3294705)

Robidoux, C., & Hewett, B. L. (2010). Collaborating virtually to develop this book: A discussion. In B. L. Hewett & C. Robidoux (Eds.), *Virtual collaborative writing in the workplace: Computer-mediated communication technologies and processes* (pp. 402–435). Hershey, PA: IGI Global.

Schendel, E., Neal, M., & Hartley, C. (2004). Toward a theory of online collaboration. In B. Huot, B. Stroble, & C. Bazerman (Eds.), *Multiple literacies for the 21st century* (pp. 195–209). Cresskill, NY: Hampton.

Selfe, C. (1992). Computer-based conversations and the changing nature of collaboration. In J. Forman (Ed.), *New visions of collaborative writing* (pp. 147–169). Portsmouth, NH: Boynton-Cook.

Sharples, M., Goodlet, J., Beck, E. E., Wood, C. C., Easterbrook, S., & Plowman, L. (1993). Research issues in the study of computer supported collaborative writing.

In M. Sharples (Ed.), *Computer supported collaborative writing* (pp. 9–28). London, UK: Springer-Verlag.

Speck, B., Dice, C. P., Johnson, T. R., & Heaton, L. B. (2008). *Collaborative writing: An annotated bibliography*. Westport, CT: Greenwood.

Staples, K., & Ornatowski, C. M. (1997). *Foundations for teaching technical communication: Theory, practice, and program design*. Greenwich, CT: Ablex.

Tovey, J. (2001). Building connections between industry and university: Implementing an internship program at a regional university. *Technical Communication Quarterly, 10*(2), 255–239.

Weaver, R. (1985). *The ethics of rhetoric*. Davis, CA: Hermagoras.

Weinberger, D. (2007). *Everything is miscellaneous: The power of the new digital disorder*. New York, NY: Henry Holt.

Weiss, E. H. (2002). Egoless writing: Improving quality by replacing artistic impulse with engineering discipline. *Journal of Computer Documentation, 26,* 3–10.

Yates, J., Orlikowski, W., & Rennecker, J. (1997). Collaborative genres for collaboration: Genre systems in digital media. *Proceedings of the 30th Hawaii international conference on System Sciences* (pp. 50–59). New York, NY: IEEE Computing Society.

Zemliansky, P., & St.Amant, K. (2007). *Handbook of research on virtual workplaces and the new nature of business practices*. Hershey, PA: Information Science.

AFTERWORD

A Neophyte Discovers Technical Writing and Is On His Own

Dan Riordan

I began to teach technical writing in the fall of 1970. Until the spring of 1970, I had never even heard of technical writing; I was an American literature aspirant at the University of North Carolina in Chapel Hill, where I was on a departmental committee to reorganize the undergraduate English curriculum with Fred MacIntosh—one of a number of World War II veterans who had developed a commitment to clear, practical writing during their years of service and who were working to introduce courses on practical writing into the university curriculum. Fred found out that I, like almost every literature graduate that year, was having trouble finding a job, and he recommended me for a position teaching technical writing at what was then Stout State University in Menomonie, Wisconsin. I initially took the position, thinking I would stay for 2 years and move on to a career teaching literature in the Rocky Mountain states. (Instead, I retired after some 40 years of teaching in that very field.) To help me get this initial job, MacIntosh put me into his course, Expository Writing; introduced me to the genres of technical writing; taught me the basics of clear, effective writing that I used throughout my career; armed me with Houp and Pearsall's 2nd edition of *Reporting Technical Information*; and sent me off into the world.

That was all there was in those days. My notes from MacIntosh and the new text from two other World War II vets—no journals, no books, no conferences, no CPTSC, no ATTW, and for me in a rural setting, no STC. (The school at which I was teaching happened to be too far away to become a regular attending member at local conferences or other events.) Finally, CPTSC started, and even though I had no travel money to attend the conferences, I joined the organization and received each year's *Proceedings*. For me, those *Proceedings* were the equivalent of receiving the Sears Christmas catalog as a child. I pored over the abstracts, found ideas to try, and gained encouragement from others who were doing what I was trying to. The contributors to these early proceedings helped me to shape individual courses and eventually to build a program at the University of Wisconsin-Stout (which Stout State became after Wisconsin merged its state system with the University of Wisconsin system).

THE PROFESSION MATURES INTO A RESEARCH-BASED FIELD

How all that has changed. No longer do new members of the field have to enter the profession like Adam and Eve, who "with wandering steps and slow . . . took their solitarie way" (Milton, 1674) out into the world. Today, faculty teaching in technical communication programs have in their hands the work of many articulate scholars and teachers who have created a profession. Today, there exists a body of work in which educators and administrators in technical communication present investigations of how to base the endlessly fascinating world of practical communication solidly on research and theory.

One area that our profession has always dealt with (it comes, so to speak, with the territory) is how academics relate to industry. As members of an applied field, we have the problem and challenge of having our feet in both worlds, worlds whose demands and values often seem to clash. As a result, we can be seen as living in an Ivory Tower by some of our peers or as sellouts by others. Our work, however, requires that we merge the two worlds and seek to meld the advantages of the liberal arts tradition to which most of us belong. We do so either by training or by our location within our universities. In so doing, we address the practical human problems of making a living by working and communicating in our counties' economies. (In 1970, I did not know that technical communicators existed in Europe or Asia or anywhere outside the United States.)

Because this relationship between academia and industry is a perennial issue, both organizations like CPTSC and scholars in our field have returned to it regularly. And in this volume, you have the latest thinking on this all-important issue. More volumes will deal with this topic in the future, but for now you have a veritable blueprint for researching, presenting, and articulating the dual professional world we live in.

A REVIEW OF THE CONTENTS OF THIS BOOK

This collection has presented a range of topics that can help bring together the academy and industry sides of the field. Now, it's up to us—the members of the discipline—to consider how to forge the kind of partnerships noted in this text. So how do we proceed? How can we use the contents of this volume to guide our actions? In these final, few pages, I'd like to reflect upon those questions.

Topics to research and how to explore them are constant themes addressed by the authors in this collection. Robidoux and Overbey, for example, suggest rhetorical questions that need to be answered when academics and industry workers collaborate. Working from his experiences as a consultant and researcher, Bernardt extends this examination by listing pertinent topics that a consulting colleague posed to him. These questions include the best ways to plan and visualize documents, the role of schemata in the decision-making process, the impact of prior knowledge on reading behavior, and many more. Bernhardt suggests that, based on lists like his, our departments select a coordinated research agenda that focuses the work of all the members and their students on a few selected areas.

Discussion of a research agenda naturally leads to the question of how to engage effectively in such research. Bernhardt's entry in this collection offers narratives of how the University of Delaware and a number of European universities developed coordinated research focuses. Your department could do the same. Start with your strengths and follow the steps Bernhardt outlines. Throughout this volume, however, you will find examples of many types of research and research reporting. Surveys (the chapters by Smith, Spartz, Weber, and Howard), Case studies (the entries by Katz; Robidoux, Hewett, & Overbey; Howard; Heifferon; and Carpenter), and posing research questions (the aforementioned chapters by Bernhardt; Robidoux, Hewett, & Overbey) are all covered in some detail or are themselves an example of the task.

In addition to noting the research work we can do, this text is a virtual handbook of how to approach the academy-industry dynamic. Most of the authors provide insights on negotiating this dynamic, which all too often is initially regarded as antagonistic. Effective discussions of how to handle the dynamic are given by contributors such as Dubinsky; Nugent & José; Heifferon; and Robidoux, Hewett, & Overbey. Moreover, many of the chapters in this collection contain clear guidelines to help you, the reader, as you deal with similar situations. Repeatedly, you will see both the call for openness and trust as well as the caution that those qualities may take time to emerge when establishing academy-industry relationships.

In a similar vein, the contributing authors introduce and examine the "stakeholder" concept so important in our theoretical and pedagogical practices. For example, Nugent & José, Price, Dubinsky, and Carpenter, all explore connections to theory when examining academy-industry relationships and partnerships in their respective chapters. Dubinsky, in particular, leads readers through the difficulties of negotiating with people across the academic, nonacademic divide. He does so by demonstrating the value of what he calls being "native" to an area, a quality not everyone agrees a university has.

Additionally, the actual work of creating an academic-industry partnership is presented for you in excellent detail. The chapters by Price, Smith, Katz, and Howard, in particular, all present models of what to do when creating such partnerships. In examining such models, these authors include topics such as the sequencing of course design, models of ways to see the partnership, and detailed discussions of how to manage your own work flow. Katz, for instance, details her work in contacting partners and students, down to the daily level. Howard, in turn, explains a different dimension of the dynamic: how to enter into contracts to do work for industry. Such examinations are important, for I suspect this area of contractual relations in partnerships is virtually unknown to many of us. Fortunately, Howard's reporting on his work at Clemson gives us a model for framing the work and should lead us to ask relevant questions at our own universities on how to handle such ventures.

The task of coordinating internships and service-learning projects is, it seems to me, easier in a large metropolitan area, just because more companies locate there. But many academics in technical communication are situated in rural areas. Individuals working in these areas will find the articles by Carpenter and by Spartz & Weber helpful in this regard. Within this vein, Dubinsky, whose works with local communities, and Spartz & Weber, who work with entrepreneurs, expand the parameters of "academic-industrial" relationships to a much wider scope that includes a vibrant and more expansive set of communities for us and our students to interact with.

Another facet of the academic-industry interaction is evaluation. Within this context, a range of seemingly unexpected factors can affect interactions among the different parties. From my own experiences, I know that when I asked to see what my students have done on the job, I have encountered the amazing response that I could not see any of the work students performed for their internships. Why? That work was seen as confidential by the internship provider. So how to assess the work? Fortunately, Price and Smith's entries in this collection provide details for handling this often sticky situation.

If you are working with internships, the benefits to your students are almost immediately apparent. As many of the contributors to this collection note, these benefits can include the ability to merge theory and practice, the experience of crossing the boundary between academy and workplace, the thrill of creating work that others need in order to do their work, the feeling that one is contributing, and the chance to receive a job offer from the host company. But the benefits to the internship provider are sometimes not as clear. A number of hosts (as reported in Dubinsky, Smith, and Katz's chapters) note that one of the main benefits they, as internship providers, received from such relationships was the ability to help another person get started—a feeling that may come as a surprise to some. Others internship providers, in turn, report that having an intern helps keep them up-to-date. And I have direct experience with such situations. Recently, for instance, the head of a local engineering firm told me that they really wanted to hire someone "young" who understood how to use Facebook, Twitter, and websites in order to bring the company up to speed. After that meeting, I called our program internship director and alerted him to the opportunity so that he could turn this chance comment in to an opportunity for one of our students.

Career advice, both in terms of research questions and curriculum content, appears in several chapters in this text. Bernhardt lists pertinent questions about research, which, although focused on the pharmaceutical industry, can easily be used in other situations. Similarly, Smith presents results of a survey of managers that gives very explicit information—and, by the way, demonstrates how to do a small qualitative project that can have wide-ranging effects. If we had a number of such small surveys published regularly, say on the CPTSC website, we could continue to build answers to the vexing questions.

In certain cases, theory can serve as a guide for addressing such relationships and the interactions that foster and maintain them. Within the context of this collection, Heifferon's chapter in particular gives an excellent example of how to use rhetorical theory to resolve the issues caused when an outsider (read "academic") enters a nonacademic situation. As you watch her work through Geertz, Foucault, and activity theory, you see how to use these theories in a consultative situation with an industry client. Further, Heifferon's chapter reveals how educators in the field can demonstrate the value of theories to students. A similar connection to theory can be seen in Robidoux, Hewett, & Overbey's chapter, which details how the authors used theory to address the difficulties of collaborating when the groups involved have similar values but much different takes on the meaning of those values in different contexts.

Another interesting aspect of this collection is the various references readers can find within the chapters themselves (or the sources within the source). Any collection of manuscripts by scholars will present you with excellent reading lists in the various references sections. The authors of the chapters here, however, often offer specialized reading lists on topics central to understanding academy-industry relationships. Covered are theory and research on internships (Price), user concerns (Howard), activity theory (Heifferon), entrepreneurial writing (Spartz & Weber). In addition, two articles are often mentioned, Sherry Little's 1993 study of internships and Carolyn Miller's "A Humanistic Rationale for Technical Writing" (1979). In these ways, this collection of works on academy-industry interactions represents both a central and important resource on the topic as well as a relatively comprehensive (and unique) aggregation of the scholarship in this area.

IN CONCLUSION, THIS COLLECTION FOCUSES ON STUDENTS AND OPENNESS

Throughout this book, you will find theoretical and practical advice for your research, for your pedagogy, and for your interaction with this important dimension of our profession. The things I like the most about this collection are the repeated calls to benefit our students by contextualizing their learning, the repeated return to the necessity to know your strengths, and the repeated advocacy of entering into relationships openly, but with an awareness of value differences. (Such differences, after all, are openings for increased understanding and personal growth for all involved.) I missed only one thing: the role of social media in the academy-industry dynamic. That topic, I suppose, will appear in another volume. Technical communication has come great distances since I, and people of my generation, left the paradise of graduate school and entered nearly alone the mostly unknown world of technical communication. You have in your hands a guide that will make you aware of the path and the difficulties along the way, but will make sure you see the vistas and not just the stones in the path.

REFERENCES

Little, S. B. (1993). The technical communication internship: An application of experiential learning theory. *Journal of Business and Technical Communication, 7*(4), 423–451.

Miller, C. R. (1979). A humanistic rationale for technical writing. *College English, 40*(6), 610–617.

Milton, J. (1674). Ed. Luxon, Thomas H. The John Milton Reading Room. *Trustees of Dartmouth College.* Retrieved October 20, 2012, from http://www.dartmouth.edu/~milton

Editors' Biographies

Tracy Bridgeford is an associate professor of technical communication at the University of Nebraska Omaha, where she directs the Technical Communication program and the English Master's program. She is coeditor of *Programmatic Perspectives*, the journal of the Council for Program in Technical and Scientific Communication, and serves on its executive committee as information officer. She has contributed chapters to *Resources in Technical Communication: Outcomes and Approaches, Teaching Writing with Computers: An Introduction, and Innovative Approaches to Teaching Technical Communication*, which she also coedited. She has also published in *Kairos: A Journal of Rhetoric, Technology, and Pedagogy*. She coedited a special issue of *Technical Communication Quarterly*, "Techne and Technical Communication" (2002).

Kirk St.Amant is a professor of technical and professional communication and of international studies at East Carolina University, Greenville, North Carolina.

Contributors

Stephen A. Bernhardt holds the Andrew B. Kirkpatrick, Jr. Chair in Writing at the University of Delaware, in which position he promotes strong writing and communication skills across the university. He is the author of *Writer's Help*, a new, Web-based reference handbook (Bedford/St. Martin's). He teaches courses in scientific and technical communication, first-year composition, computers and writing, and grammar and style. He is past president of both the Council for Programs in Technical and Scientific Communication and the Association of Teachers of Technical Writing.

Russell Carpenter directs the Noel Studio for Academic Creativity at Eastern Kentucky University, where he is also an assistant professor of English. He has a Ph.D. in texts & technology from the University of Central Florida. Recent books include *Higher Education, Emerging Technologies, and Community Partnerships* (with Melody Bowdon; 2011), *Cases on Higher Education Spaces* (2012), and *The Routledge Reader on Writing Centers and New Media* (with

Sohui Lee; 2013). He currently serves as president of the Southeastern Writing Center Association and chair of the National Association of Communication Centers.

James Dubinsky is an associate professor of rhetoric and writing in the Department of English at Virginia Tech and executive director of the Association for Business Communication. Previously he served as founding director of Virginia Tech's Professional Writing program and as founding director of the university's Center for Student Engagement and Community Partnerships, now VT-ENGAGE. He has received college-level awards for teaching and outreach and was one of Virginia Tech's first two recipients of the Scholarship of Teaching and Learning award. His publications include *Teaching Technical Communication: Critical Issues for the Classroom* and articles in journals ranging from *Technical Communication Quarterly* to *Michigan Journal of Community Service Learning.*

Barbara Heifferon is a professor of English and writing director at Louisiana State University. She is a national leader of medical rhetoric, a subspecialty that bridges several fields. She coedited the first volume of essays on medical rhetoric, *Rhetoric of Healthcare Essays: Toward a New Disciplinary Inquiry* (2008) and was the invited author of *Writing in the Health Professions* (2002), the first medical rhetoric text in the well-known Allyn & Bacon Series in Technical Communication. In addition to coediting a special issue of *Technical Communication Quarterly*, the first on medical rhetoric (2000), Heifferon has published 16 articles.

Beth L. Hewett, Educational Consultant, Defend & Publish, LLC, is a leading expert in online writing instruction (OWI). She chaired/co-chaired the Conference on College Composition and Communication (CCCC) committee for Effective Practices in Online Writing Instruction for seven years. She was the initial OWI writing program developer for Smarthinking, Inc., and redesign consultant for TutorVista. She coedited at *Kairos* for seven years. Hewett's publications include *Reading to Learn and Writing to Teach: Literacy Strategies for OWI* (Bedford, forthcoming) and *The Online Writing Conference: A Guide for Teachers and Tutors* (Heinemann, 2010; updated, Bedford, forthcoming). She also coedited/coauthored *Foundational Principles of OWI* (Parlor Press, forthcoming), *Virtual Collaborative Writing in the Workplace: Computer-Mediated Communication Technologies and Processes* (IGI Global, 2010), *Preparing Educators for Online Writing Instruction: Principles and Processes* (NCTE, 2004), and *Technology and English Studies: Innovative Career Paths* (LEA, 2005).

Tharon Howard is author of *Design to Thrive: Creating Social Networks and Online Communities That Last* and four other books. He conducts sponsored research for clients including IBM, NCR Corp., AT&T, and Pearson Publishing. He is community manager for UTEST, a 21-year-old online community for usability testing professionals, and has received the UXPA's "Extraordinary

Service Award" and the STC's "Usability SIG Achievement Award". Howard teaches in the doctoral and master's degree programs at Clemson University. He also directs Clemson's Multimedia Authoring Teaching and Research Facility, where faculty and graduate students design interactive, multimodal productions and experiment with emerging instructional technologies and interface designs.

Laurence José is an assistant professor of writing at Grand Valley State University, where she teaches courses in professional writing, document design, style, and genre theory. Her research interests include international communication, technical communication pedagogy, literacy studies, and linguistics. She currently serves as the associate editor of *Programmatic Perspectives*, the journal of the Council for Program in Technical and Scientific Communication. José has published in technical communication and linguistics journals. Her most recent work is a forthcoming book chapter on globalizing technical communication programs.

Susan Katz is an associate professor of English; she spent 12 years in television and advertising before turning to the study of writing in public and private organizations. She earned her Ph.D. in communication and rhetoric at Rensselaer Polytechnic Institute. She teaches graduate and undergraduate courses in the rhetoric of science and technology at North Carolina Sate University and coordinates the English Department Internship. Katz is the author of *The Dynamics of Writing Review: Opportunities for Growth and Change in the Workplace* and *Start Your Career: 5 Steps to Finding the Right Job after College* and coauthor (with Lee Odell) of *Writing Now.*

Jim Nugent is an assistant professor of writing and rhetoric at Oakland University. He has a Ph.D. in rhetoric and technical communication from Michigan Technological University, a master's degree in technical writing from Illinois State University, and a bachelor's degree in math and computer science from the University of Illinois at Urbana-Champaign. His research interests include neosophistic rhetorical theory, the teaching of technical writing, and certificate programs in technical communication.

Kenneth Price teaches professional and technical communication courses at the University of Wisconsin–River Falls. He was formerly a member of the graduate faculty at Missouri State University; California State University, Chico—where he directed the technical writing certificate program; University of Alaska Anchorage; and Western Carolina University. He also served as a software documentation consultant to Macromedia. He regularly teaches courses in authoring digital media, computer documentation, scientific and technical editing, as well as a number of specialized courses in technical communication.

Charlotte Robidoux is a business consultant at Hewlett-Packard Company and has over 20 years of experience in technical communication. At HP, she oversees the single sourcing strategy and implementation for Enterprise Services. She has a Ph.D. in rhetoric and technical communication from the Catholic University of America. Robidoux is the author of "Rhetorically Structured

Content: Developing a Collaborative Single-Sourcing Curriculum" (*Technical Communication Quarterly*), coauthor of "Is There a Write Way to Collaborate?" (*Intercom*), and coauthor of "Streamline Your Path to Metadata" (in *The Information Management & Architecture Framework*, October 2009). She is coeditor of *Collaborative Writing in Virtual Workplaces: Computer-Mediated Communication Technologies and Tools* (2010).

Carolyn Rude began teaching technical communication at Texas Tech in 1981 and retired from Virginia Tech in 2013. During that time she observed, with regret, the widening gap between academics and industry. She is a past president of the Association of Teachers of Technical Writing and a Fellow of both ATTW and the Society for Technical Communication. She is the author of the textbook *Technical Editing* and of articles on environmental rhetoric, genre, and decision making.

Herb J. Smith is a professor of English, technical communication, and media arts at Southern Polytechnic State University, Marietta, Georgia. He currently serves as coordinator of the B.S. degree in technical communication and coordinates internships for majors in all three of the department's undergraduate programs. Smith also administers the mentorship program for Project Portfolio, the department's senior capstone course. He has taught technical and professional communication for more than 30 years, including a semester as Gastprofessur of Technical Writing at Fachhochschule Merseberg, Germany. Smith has published widely in most of the major technical communication journals. *Portfolios for Technical and Professional Communicators* (coauthored with Haimes-Korn) provides detailed information about paper and electronic portfolios for technical and professional communicators.

John M. Spartz is an assistant professor of professional/technical communication and emerging media in the Department of English and Philosophy at the University of Wisconsin–Stout. His research draws connections between the fields of professional/technical communication and entrepreneurship, with a focus on the disciplinary and pedagogical import of that relationship. His work has appeared in *The Writing Instructor* and *The Writing Center Journal* and is scheduled to appear in *Programmatic Perspectives* and the *Journal of Business and Technical Communication.*

Ryan Weber is an assistant professor of English and director of business and technical writing at the University of Alabama in Huntsville. His work has appeared in *Journal of Advanced Composition*, *College Composition and Communication*, and *Journal of Business and Technical Writing*.

Index